Audel™

Electrical Course for Apprentices and Journeymen

All New Fourth Edition

Audel™
Electrical Course for Apprentices and Journeymen

All New Fourth Edition

Paul Rosenberg

WILEY

Wiley Publishing, Inc.

Vice President and Executive Group Publisher: Richard Swadley
Vice President and Executive Publisher: Robert Ipsen
Vice President and Publisher: Joseph B. Wikert
Executive Editorial Director: Mary Bednarek
Editorial Manager: Kathryn A. Malm
Executive Editor: Carol A. Long
Senior Production Editor: Fred Bernardi
Development Editor: Regina Brooks
Production Editor: Pamela Hanley
Text Design & Composition: Wiley Composition Services

Library of Congress Cataloging-in-Publication Data:

ISBN: 0-764-54200-1

Printed in the United States of America

10 9 8 7 6 5 4 3 2 1

Contents

Introduction

An apprentice electrician prepares to become a journeyman typically during a four-year period. These four years include 2000 hours per year of on-the-job training, or a total of 8000 hours. During off-hours an apprentice studies electrical theory, methods, equipments, and the *NEC*.

My purpose in writing this book was to provide the apprentice or journeyman with most of the information he or she is required to know. I have drawn on my experience as a former instructor of apprentice and journeymen electricians to include most of the vital material on both electrical theory and its applications.

This book has been planned as a study course either for the classroom or as a self-taught program. It may be utilized without any other books on electrical theory.

Very little on the *NEC* is included since two other Audel books offer abundant information on it. *Guide to the National Electrical Code,* which is updated annually as the *NEC* is changed, gives a very complete interpretation of the *Code. Questions and Answer for Electrician's Examinations* can further help the electrician toward a thorough knowledge of the *NEC*.

Trigonometry is covered briefly in this book, because it is useful in making mathematical calculations of alternating currents. For the reader who is not familiar with trigonometry, there are other means of explanation.

It is not the intent of this book to give a complete discussion of *all* electrical subjects. However, with the basic information presented here, the apprentice or journeyman can gain an understanding of operational theory and progress even further, if he or she wishes.

I sincerely hope that this book will be of value to you, the electrician. It has been my good fortune to learn a great deal from others in our field, and I have presented here the information I have gained. Any knowledge that you or future electricians gain from this book will make my time spent in writing it worthwhile.

The basics of electricity really do not change, but the applications of these basics do change. Therefore, I hope that you will continue your studies throughout your career and keep abreast of the continual changes in the field. You will find that in modern society the person with the know-how is the person who advances.

A college degree is a valuable asset—get one if you can. But remember that much of the information offered by a degree program may be gained by self-study. Many people with technical know-how are needed to back up the engineering profession, and a technical education is receiving increased recognition.

I wish to extend my sincere thanks to the many fine people I've worked with through the years. Your contributions have been critical.

Paul Rosenberg

Audel™

Electrical Course for Apprentices and Journeymen

All New Fourth Edition

Chapter 1

Electricity and Matter

Electricity is one of the great wonder-workers of our modern world. It is a force that powers thousands of inventions that make life more pleasant. Electricity is a property of certain particles to possess a force that can be used for the transmission of energy. Whenever electricity is used, you may be assured that an equal amount of some other form of energy was previously used to produce the electricity.

In order to gain an understanding of what electricity is, we must go into some study of matter, molecules, atoms, and elements. This is what may be termed the *electron theory.*

A Greek philosopher, Thales, in about 600 B.C., discovered that a piece of amber rubbed with a woolen cloth would attract pieces of chaff and other light objects, much as a magnet attracts iron filings. The Greek word for amber is *elektron* and it probably is from this word that the English words "electricity" and "electron" were derived. More on this phenomenon will be covered later.

Elements, Atoms, Molecules, and Compounds

All substances may be termed matter, and matter may be liquid, solid, or gaseous. A good example is water. Water may be a solid (ice), a liquid (water) with which we wash or drink, and a gas or steam (vapor), which we get when water is boiled. Whether it is ice, liquid, or vapor, its chemical makeup does not change; only the state in which it appears changes.

Elements are substances that can't be changed, decomposed by ordinary types of chemical change, or made by chemical union. There are over 100 known elements, distinguishable by their chemical and physical differences. Some common elements are copper, silver, gold, oxygen, hydrogen, sulfur, zinc, lead, helium, and uranium.

A *molecule* is the smallest unit quantity of matter that can exist by itself and retain all the properties of the original substance. It consists of one or more atoms.

Atoms are regarded as the smallest particles that retain the properties of the element and which, by chemical means, matter may be divided into.

Some of the more than 100 elements and their characteristics are given in Table 1-1. From this table and the symbols for the elements appearing in this table, it will be easier to gain insight concerning compounds. Some everyday compounds are

Water (H_2O): Two atoms of hydrogen and one atom of oxygen.

Sulfuric acid (H_2SO_4): Two atoms of hydrogen, one atom of sulfur, and four atoms of oxygen.

Salt (NaCl): One atom of sodium and one atom of chlorine.

Table 1-1 Elements and Their Characteristics

Atomic Number	Element	Symbol	Atomic Weight
13	Aluminum	Al	26.98
51	Antimony	Sb	121.76
18	Argon	A or Ar	39.948
56	Barium	Ba	137.34
4	Beryllium	Be	9.01
83	Bismuth	Bi	208.98
5	Boron	B	10.81
48	Cadmium	Cd	112.40
20	Calcium	Ca	40.08
6	Carbon	C	12.011
55	Cesium	Cs	132.905
17	Chlorine	Cl	35.453
24	Chromium	Cr	51.996
27	Cobalt	Co	58.93
29	Copper	Cu	63.54
9	Fluorine	F	19.00
79	Gold	Au	196.967
2	Helium	He	4.003
1	Hydrogen	H	1.008
26	Iron	Fe	55.847
82	Lead	Pb	207.21
3	Lithium	Li	6.94
12	Magnesium	Mg	24.32
25	Manganese	Mn	54.94
80	Mercury	Hg	200.61

(continued)

Table 1-1 (continued)

Atomic Number	Element	Symbol	Atomic Weight
42	Molybdenum	Mo	95.94
10	Neon	Ne	20.183
28	Nickel	Ni	58.71
7	Nitrogen	N	14.007
8	Oxygen	O	16.000
15	Phosphorus	P	30.974
78	Platinum	Pt	195.09
19	Potassium	K	39.102
88	Radium	Ra	226.05
45	Rhodium	Rh	102.91
34	Selenium	Se	78.96
14	Silicon	Si	28.09
47	Silver	Ag	107.87
11	Sodium	Na	22.991
38	Strontium	Sr	87.62
16	Sulfur	S	32.066
90	Thorium	Th	232.038
50	Tin	Sn	118.69
74	Tungsten	W	183.85
92	Uranium	U	238.03
30	Zinc	Zn	65.37

Some forms of matter are merely mixtures of various elements and compounds. Air is an example; it has oxygen, nitrogen, helium, argon, neon, and some compounds such as carbon dioxide (CO_2) and carbon monoxide (CO).

One may wonder what all of this has to do with electricity, but it is leading up to an explanation of the electron theory, which follows.

Electron Theory

An atom may be roughly compared to a solar system in which a sun is the nucleus around which orbit one or more planets, the number of which depends on which atom we pick from the various elements. (Bear in mind that this is not a completely accurate description, as electrons seem to move in figure eights, rather than in circles. Nonetheless, the comparison between a solar system and an atom is useful.)

The nucleus is composed of protons and neutrons, and orbiting around this nucleus of protons and neutrons are electrons. An *electron* is a very small negatively charged particle. Electrons appear to be uniform in mass and charge and are one of the basic parts of which an atom is composed. The charge of the electron is accepted as 4.80×10^{-10} absolute electrostatic unit. This indicates that all electrons are alike regardless of the element of which they are a part.

A *neutron* is an elementary particle with approximately the mass of a hydrogen atom but without an electrical charge.

A *proton* is an elementary particle having a positive charge equivalent to the negative charge of an electron but possessing a mass approximately 1845 times as great.

From Table 1-1, we find the *atomic number* (number of protons in the nucleus) of hydrogen is 1, helium is 2, lithium is 3, beryllium is 4, etc. Figure 1-1 shows the atoms of hydrogen, helium, lithium, and beryllium, with the electrons orbiting around the nucleus of neutrons and positively charged protons. Notice that the positive charge of the protons in the nucleus equals the negative charge of the electrons and holds them in orbit.

Electrons may be released from their atoms by various means. Some atoms of certain elements release their electrons more readily than atoms of other elements. If an atom has an equal number of electrons and protons, it is said to be in balance. If an atom has given up some of its electrons, the atom will then have a positive charge, and the matter that received the electrons from the atom will be negatively charged. Some external force must be used to transfer the electrons.

Before progressing further, any electrical discussion must include *static electricity,* for a better understanding of insulation and conductors, as well as to carry on with the discussions of dislodging electrons. The word "static" means at rest. There are some applications where static electricity is put to use, but in other cases it is detrimental and must be avoided. We are faced with lightning, which is static electricity discharges attempting to neutralize opposite charges. Since we have to live with lightning's harmful effects, we should know how to cope with it. The methods of avoiding the harmful effects of lightning are not fully discovered but much progress has been made.

One method of dislodging electrons is by the friction of rubbing a hard rubber rod with a piece of fur. The fur will give up some

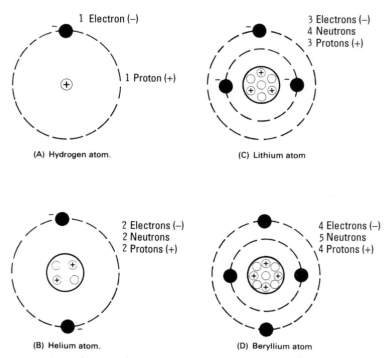

Figure 1-1   Atoms: electrons, neutrons, and protons. Electrons have a negative (–) charge, protons have a positive (+) charge, and neutrons are neutral.

electrons to the hard rubber rod, leaving the fur with a positive charge, and the hard rubber rod will gain a negative charge. Then, again, a glass rod rubbed with silk will give up electrons to the silk, making the silk negatively charged and leaving the glass rod positively charged.

What actually transpires is that the intimate contact between the two surfaces results in the fur being robbed of some of its negative electrons, thereby leaving it positively charged, while the rubber rod acquires a surplus of negative electrons and is thereby negatively charged. It is important to note that this surplus of negative electrons doesn't come from the atomic structure of the fur itself. It is found that, in addition to the electrons involved in the structure of materials, there are also vast numbers of electrons "at large." It is from this source that the rubber rod draws its negative charge of electrons.

If a hollow brass sphere is supported by a silk thread as in Figure 1-2 (silk is an insulator), and a hard rubber rod that has received a negative charge, as previously described, is touched to the brass sphere, the brass sphere will also be charged negatively by a transfer of electrons from the rod to the ball. The ball will remain negatively charged as it is supported by the insulating silk thread.

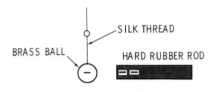

Figure 1-2 A negatively charged hard rubber rod touched to a hollow brass ball supported by a silk thread will negatively charge the brass ball.

Now if the same experiment is tried with the hollow brass sphere supported from a metal plate by a wire, the rubber rod will transfer electrons to the ball but the electrons will continue through the wire and metal plate and eventually to earth (see Figure 1-3).

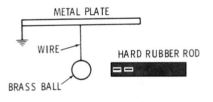

Figure 1-3 When a negatively charged hard rubber rod is touched to a hollow brass ball supported from a metal plate by a wire, the negative charge will move through the metal wire and on to earth.

When a body acquires an electrical charge as, for example, the hard rubber rod or the glass rod previously described, it is customary to say that the lines of force emanate from the surface of the electrified body. By definition, a line of electrical force is an imaginary line in space along which electrical force acts. The space occupied by these lines in the immediate vicinity of an electrified body is called an *electrostatic field of force* or an *electrostatic field*.

In Figure 1-2, the hollow ball was negatively charged and the lines of force emanated from it or converged on it in all directions (see Figure 1-4).

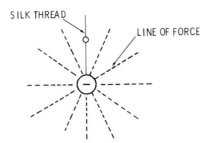

SILK THREAD

LINE OF FORCE

Figure 1-4 Lines of force from an electrically charged hollow ball emanate in, or converge from, all directions.

Static electrical charges may be detected by an *electroscope*. The simplest form of an electroscope is a light wooden needle mounted on a pivot so that it may turn about freely. A feather or a pith ball suspended by silk thread may also be employed for the purpose.

The electroscope most used was devised by Bennett and consists of a glass jar (Figure 1-5) with the mouth of the jar closed by a cork. A metal rod with a metal ball on one end (outside the jar) and a stirrup on the other passes through the cork, and a piece of gold leaf is hung over the stirrup so that the ends drop down on both sides.

When an electrified rod is brought close to the hollow brass ball, the electrostatic field charges the ball. In Figure 1-5, the rod is

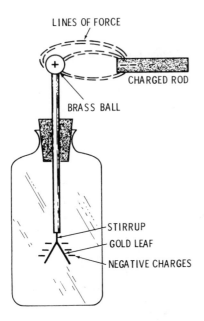

LINES OF FORCE

CHARGED ROD

BRASS BALL

STIRRUP

GOLD LEAF

NEGATIVE CHARGES

Figure 1-5 Gold-leaf electroscope.

negatively charged, so the electrons in the ball are repelled and the ball becomes positively charged. The electrons that were repelled from the ball go to the gold leaf, charging both halves of the gold leaf negatively, and the leaves fly apart, as illustrated in Figure 1-5.

Like charges repel each other and unlike charges attract. Since both halves of the gold leaf are charged the same, they repel. Remember that we have not touched the rod to the ball in this experiment; the electrostatic charges are transmitted by *induction.*

If a positively electrified ball (A in Figure 1-6) mounted on an insulated support is brought near an uncharged insulated body (B-C), the positive charge on ball A will induce a negative charge at point B and a positive charge at point C. If pith balls are mounted on wire and suspended by cotton threads, as shown, the presence of these charges will be manifested. The pith ball (D), electrified by contact with B, acquires a negative charge. It will be repelled by B and attracted toward A and stands off at some distance. The ball (E) is charged by contact positively and will be repelled from C a lesser distance because there is no opposite charge in the vicinity to attract it, while ball F at the center of the body will remain in its original position, indicating the absence of any charge at this point. This again shows electrostatic induction. The electric strain has been transmitted through the intervening air (G) between A and B and reappears at point C.

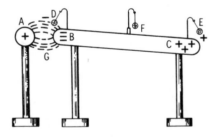

Figure 1-6 Illustration of charges produced by electrostatic induction.

In Figure 1-6, the air in the space (G) between A and B is called a *dielectric.* The definition of a dielectric is any substance that permits induction to take place through its mass. All dielectrics are insulators, although the dielectric and insulating properties of a substance are not directly related. A dielectric is simply a transmitter of a strain.

When a dielectric is subjected to electrostatic charges, the charge tries to dislodge the electrons of the atoms of which the dielectric is composed. If the stress is great enough, the dielectric will break down and there will be an arc-over. Dielectrics play a very important role in the theory of the electrical field.

Electric Current

We learned earlier that static electricity refers to electrical charges that are stationary—that is to say, a surplus of electrons, or the lack of same, that stay in one place, not in motion.

Electrons in motion constitute an *electric current*. Thus, if electrical pressure from a battery, generator, or other source is applied to an electrical conductor, such as a copper wire, and the circuit is closed, electrons will be moved along the wire from negative to positive. These electrons pass from atom to atom and produce current. The electrons that move are free electrons. They may be compared to dominoes set on end. If the first one is pushed over, it knocks the next one over and so on. This progression of movement of energy occurs at the speed of light, or approximately 186,000 miles per second.

During the early days of electrical science, electricity was considered as flowing from positive to negative. This is opposite to the electron theory. While in the study of this course the direction of flow might seem irrelevant, in electronic circuits the proper direction of flow is very important. Therefore, in our studies we will use the right direction of flow, namely, negative to positive in line with the electron theory.

There are basically three forms of electrical current, namely (1) direct current (DC), (2) pulsating direct current (pulsating DC), and (3) alternating current (AC).

Figure 1-7 compares the flow of water to DC. Pump *A* may be compared to a battery or a generator driven by some external force,

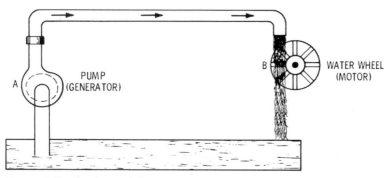

Figure 1-7   Analogy of direct current.

and wheel *B* may be compared to a DC motor, with the current flowing steadily in the direction represented by the arrows. This may also be represented as in Figure 1-8.

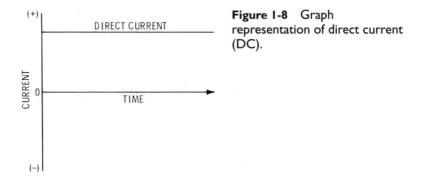

Figure 1-8   Graph representation of direct current (DC).

Now, if generator *A* in Figure 1-7 were alternately slowed down and speeded up, the current would be under more pressure when the pump was speeded up and less pressure when the pump slowed down, so the water flow would pulsate in the same direction as represented in Figure 1-9. It would always be flowing in the same direction, but in different quantities.

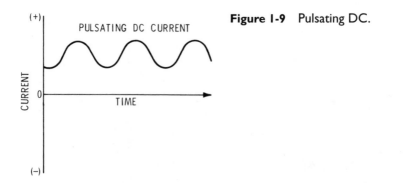

Figure 1-9   Pulsating DC.

In Figure 1-10 we find a piston pump (*A*) alternately stroking back and forth and thus driving piston *B* in both directions alternately. Thus, the water in pipes *C* and *D* flows first in one direction and then the other. Figure 1-11 illustrates the flow of AC; more will be covered later.

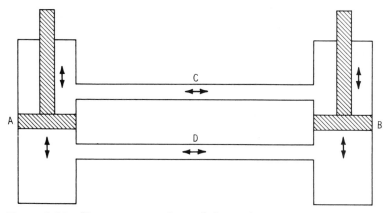

Figure 1-10   Piston pump analogy of alternating current.

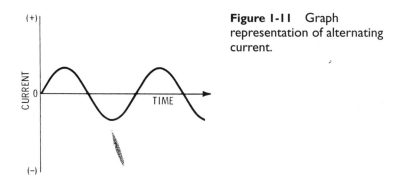

Figure 1-11   Graph representation of alternating current.

Insulators and Conductors

An insulator opposes the flow of electricity through it, whereas a conductor permits the flow of electricity through it. It is recognized that there is no perfect insulator. Pure water is an insulator, but the slightest impurities added to water make it a conductor. Glass, mica, rubber, dry silk, etc., are insulators, while metals are conductors.

Although silver is not exactly a 100 percent conductor of electricity, it is the best conductor known and is used as a basis for the comparison of the conducting properties of other metals, so we will call its conductivity 100 percent.

Some metals are listed here in the order of their conductivity:

Silver	100%	Iron	16%
Copper	98%	Lead	15%
Gold	78%	Tin	9%
Aluminum	61%	Nickel	7%
Zinc	30%	Mercury	1%

Questions

1. What is a neutron?
2. What is a proton?
3. What is an electron?
4. Sketch a boron atom and label its parts.
5. Two pith balls are negatively charged and supported by a dry silk thread. Draw a sketch showing their relative positions when they are brought close to each other.
6. Like charges (electrical) and unlike charges (electrical). Differencs?
7. What is static electricity?
8. What is electrical current?
9. What is a perfect insulator composed of?
10. Describe and draw an electroscope.
11. What is direct current? Illustrate.
12. What is pulsating direct current? Illustrate.
13. What is alternating current? Illustrate.

Chapter 2

Units and Definitions

We are all familiar with our American (English) system of measurements, but there is a very definite trend toward the establishment of an international system based on the metric system. Actually, the metric system is less complicated than our system because all quantities are in units, tens, hundreds, thousands, etc. The metric system is not only used in the vast majority of the world, but it is also used in almost all scientific applications. Get used to the metric system now. In most ways, it is a superior system.

Fundamental and Derived Units

Some of the fundamental and derived units with which we will be dealing will be covered here. We will use some of the metric system, but the English system will also be used. We will attempt to stay with common terms and expressions with which we are familiar, but it is also necessary to become familiar with the metric system.

All physical quantities, such as force, velocity, mass, etc., can be expressed in terms of three fundamental units. These are

1. *Centimeter*: The unit of length
2. *Gram*: The unit of mass
3. *Second*: The unit of time

These constitute the basis of what is called the cgs, or "centimeter-gram-second" system of units.

Units of length have the following conversions:

$$1 \text{ centimeter (cm)} = 0.3937 \text{ inch (in.)}$$
$$1 \text{ centimeter (cm)} = 1/100 \text{ of a meter (m)}$$
$$1 \text{ millimeter (mm)} = 1/1000 \text{ of a meter (m)}$$
$$1 \text{ meter (m)} = 39.37 \text{ inches (in.)}$$
$$1 \text{ inch (in.)} = 2.54 \text{ centimeters (cm)}$$

The gram is a unit of mass. It is a measure of the amount of matter that a body contains. There is a distinction to be made between mass and weight: *Weight* refers to the force with which the earth's surface attracts a given mass. Therefore, the attraction at the earth's surface for a given mass may be expressed in pounds. On this basis, one gram is equal to 1/453.6 pound. The symbol for a gram is g.

The second is 1/60 of a minute; the symbol for the second is s.

The *electrostatic unit* (esu) of a quantity of electricity refers to a point charge that when placed at a 1-centimeter distance in air from a similar and equal charge repels it with a force of 1 dyne. To convert a number of such units to coulombs, which are the practical units, divide the total number of esu by 3×10^9.

The number 10^9 (pronounced "ten to the ninth power") is the same as 1,000,000,000, but is much easier to express. This is a system of notation used to express large quantities in a condensed form. Only the significant figures are put down, the ciphers at the end being indicated by the superscript written slightly above and to the right. Thus,

$$10^2 = 100$$
$$10^3 = 1000$$
$$10^4 = 10,000$$
$$10^5 = 100,000$$
$$10^6 = 1,000,000$$
$$10^7 = 10,000,000$$

Fractions with unity numerator and a power of 10 as denominator may be expressed by negative integers written as exponents of 10. Thus.

$$1/100 = 10^{-2}$$
$$1/1000 = 10^{-3}$$
$$1/10,000 = 10^{-4}$$
$$1/1,000,00 = 10^{-5}$$
$$1/1,000,000 = 10^{-6}$$
$$1/1,000,000 = 10^{-7}$$

The resistance of air is about 10^{26} times that of copper. If this is expressed with ciphers, it is necessary to say that the resistance of air is equal to 100,000,000,000,000,000,000,000,000 times that of copper. You may readily observe that 10^{26} is a much more convenient expression than to use a 1 and 26 ciphers after it.

In expressing the fractional parts of units or multiples of units involved, certain prefixes are used:

The prefix *micro* means 1/1,000,000 part of the quantity. A microfarad is therefore 1/1,000,000 of a farad, or 10^{-6} farad.

The prefix *milli* means 1/1000 part of the quantity referred to. A milliampere is 1/1000 of an ampere, or 10^{-3} ampere.

The prefix *centi* means 1/100 part of the unit referred to. Thus, a centimeter is 1/100 of a meter or 0.3937 of an inch, since a meter is 39.37 inches. Hence a centimeter is $1/10^{-2}$ meter.

The prefix *mega* means 1,000,000 times the unit referred to. Thus 1 megohm is equal to one million ohms, or 10^6 ohms.

The prefix *kilo* means 1000 times the unit referred to. Thus a kilowatt equals 1000 watts, or 10^3 watts.

The prefix *hecto* (which we won't refer to much) means 100 times the unit to which it refers. Thus a hectowatt is equal to 100 watts, or 10^2 watts.

Definitions

A number of definitions will be given at this point in the course. There will be others given as we progress. The reason for giving these here is that we may use electrical terminology as we progress and keep the explanations to a minimum.

Insulation: A material that by virtue of its structure opposes the free flow of current through it. Commonly used insulating materials are asbestos, ceramics, glass, mica, plastics, porcelain, rubber, and paper.

Conductor: A material that allows the free flow or passage of an electric current through its structure; generally, any wire, cable, or bus suitable for carrying electrical current.

Ampere (A): The unit of intensity of electrical current (I); rate of flow of electric charge. One ampere will deposit silver in an electrolytic cell at the rate of 0.001118 gram per second.

Ohm (Ω): The unit of resistance (R) to an electrical current; a column of mercury 106.3 cm long and having a mass of 14.4521 grams (approximately) with a 1 square millimeter cross section at 0° Celsius has a resistance of 1 ohm.

Volt (V): The unit of electrical pressure (E); electromotive force (emf); potential difference. The amount of electrical pressure required to force 1 ampere through 1 ohm of resistance.

Coulomb (C): The quantity of charge that passes any point in an electric circuit in 1 second when 1 ampere of current is present.

Watt (W): The electrical unit of energy; rate of doing work (P). The product of the applied volts and the current in the circuit: 1 ampere $\times$ 1 volt = 1 watt.

Kilowatt (kW): One thousand watts.

Kilowatt-hour: One watt for 1000 hours; or 1000 watts for one hour; or 100 watts for 10 hours, etc. Unit for recording electrical power use.

Energy: The ability to do work. Energy can be neither created nor destroyed; it is a conserved quantity. It can, however, be converted from one form to another.

Foot-pound: Unit for measuring work. It is the energy required to move a weight of 1 pound through a distance of 1 foot.

Joule (J): The unit of work (W): force acting through distance. One ampere $\times$ 1 volt $\times$ 1 second = 1 joule. One watt $\times$ 1 second = 1 joule. One coulomb $\times$ 1 volt = 1 joule.

Farad (F): The unit of capacitance (C). A capacitor has a capacitance of 1 farad when one coulomb delivered to it will raise its potential 1 volt. The farad is an impractically large quantity, so you will hear more of microfarads, or 1/1,000,000 farad (10^{-6} farad).

Henry (H): The unit of electromagnetic induction. A circuit possesses an inductance of 1 henry when a rate of current variation of 1 ampere per second causes the generation therein of 1 volt.

Megawatt (MW): 1,000,000 watts; 10^6 watts.

Volt-amperes (VA): A term used to describe alternating current; since we usually have opposition to the change of direction of current in an alternating-current circuit, the volts and amperes are very commonly out of phase. (This will be explained later.)

Kilovolt-amperes (kVA): One thousand volt-amperes.

Power Factor (PF): The phase displacement of volts and amperes in an AC circuit due to capacitance and/or inductance.

The cosine of the angle of lag or lead between the alternating current and voltage waves. (This will be explained more fully later.)

Hertz (Hz): The new name for a cycle per second.

Alternation: One-half of a cycle. See Figure 2-1.

Frequency (of AC current): The number of hertz completed.

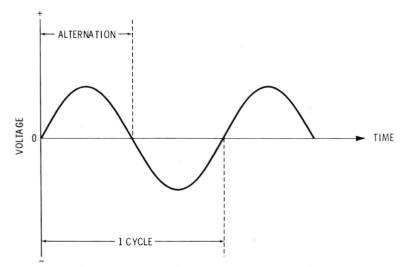

Figure 2-1   Representation of one alternation and one cycle, which consists of two alternations.

Magnetic Units

The following definitions of magnetic units are given mainly for later reference.

Gauss (G): Unit of magnetic flux density, equal to one line of magnetic flux (maxwell) per square centimeter.

Maxwell (Mx): Unit of magnetic flux, one magnetic line of flux.

Ampere-turn (At): The magnetomotive force produced by a coil, derived by multiplying the number of turns of wire in the coil by the current in amperes through it.

Oersted (Oe): Unit of magnetizing force equal to $1000/4\pi$ ampere-turns per meter.

Permeability (μ): Expresses the ratio of magnetic flux density produced in a magnetic substance to the magnetic field intensity that occasions it.

Temperature Units

Celsius (°C): A temperature scale, formerly termed centigrade and used extensively in electrical work and in the metric system.

Fahrenheit (°F): A temperature scale commonly used under our system of temperature recording.

On the Celsius scale, 0°C is the temperature at which water freezes, and 100°C is the temperature at which water boils. Both of these refer to sea level, or a barometric pressure of approximately 14.7 pounds per square inch.

On the Fahrenheit scale, water freezes at 32°F and boils at 212°F. These, as with the Celsius temperature scale, are at sea level.

It is easy to convert from Celsius to Fahrenheit and from Fahrenheit to Celsius. Examples:

To convert 100°C to Fahrenheit, take 9/5 of 100 and add 32; thus 9/5 of 100 = 180, and 180 + 32 = 212°F.

To convert 212°F to Celsius, take 212 − 32 = 180, and take 5/9 of 180 = 100°C.

In electrical trade, much electrical equipment is rated in Celsius temperature (°C), but in some cases the equipment may be rated in Fahrenheit (°F), or in a combination of °C and °F.

Questions

1. How many inches are in a meter?
2. How many centimeters are in an inch?
3. How many seconds are in a minute?
4. Write 10^{10} in common terms.
5. Write 1,000,000,000 using superscripts.
6. Write 1/1,000,000,000 using negative superscripts.
7. Define micro.
8. Define kilo.
9. Define mega.
10. Define milli.

11. Define centi.

12. What is a volt?

13. What is an ampere?

14. What is an ohm?

15. What is a watt?

16. What is a farad?

17. What is a henry?

18. What is a kilowatt?

19. What is a megawatt?

20. What is a volt-ampere?

21. What is a kilovolt-ampere?

22. What is a hertz?

23. What is a frequency?

24. Convert 20°C to Fahrenheit.

25. Convert 60°F to Celsius.

26. What was the previous name for the Celsius scale?

27. What is the barometric pressure at sea level?

Chapter 3

Electrical Symbols

Throughout this course and all through your career as an electrician, you will have need for "electrical symbols." They will be used in schematics, drawings, and numerous other places. This chapter is inserted at this point so that you might familiarize yourself with the meaning of these symbols and know where to look to find the meaning when you come across symbols.

The total coverage of electrical and electronic symbols may be obtained from American National Standards Institute, Inc., 25 W. 43rd St., New York, N.Y. 10036.

Try to familiarize yourself with the following symbols so that you will recognize them, and thus avoid rereading many explanations in this text.

Graphic Electrical Wiring Symbols

Compiled by American Standard Graphic, electrical wiring symbols for architectural and electrical layout drawings.

0.1 Drafting Practices Applicable to Graphic Electrical Wiring Symbols

A. Electrical layouts should be drawn to an appropriate scale or figure dimensions noted. They should be made on drawing sheets separate from the architectural or structural drawings or on the drawing sheets for mechanical or other facilities.

Clearness of drawings is often reduced when all different electric systems to be installed in the same building area are laid out on the same drawing sheet. Clearness is further reduced when an extremely small drawing scale is used. Under these circumstances, each or certain of the different systems should be laid out on separate drawing sheets. For example, it may be better to show signal system outlets and circuits on drawings separate from the lighting and power branch circuit wiring.

B. Outlet and equipment locations with respect to the building should be shown as accurately as possible on the electrical drawing sheets to reduce reference to architectural drawings. Where extremely accurate final location of outlets and equipment is required, figure dimensions should be noted on the drawings. Circuit and feeder run lines should be so drawn as to show their installed location in relation to the building insofar as it is practical to do so. The number and size of conductors in the runs should be identified by notation when the circuit-run symbol doesn't identify them.

C. All branch circuits, control circuits, and signal system circuits should be laid out in complete detail on the electrical drawings, including identification of the number, size, and type of all conductors.

D. Electric wiring required in conjunction with such mechanical facilities as heating, ventilating and air-conditioning equipment, machinery, and processing equipment should be included in detail in the electrical layout insofar as possible when its installation will be required under the electrical contract. This is desirable to make reference to mechanical drawings unnecessary and to avoid confusion as to responsibility for the installation of the work.

E. A complete electrical layout should include at least the following on one or more drawings:

1. Floor plan layout to scale of all outlet and equipment locations and wiring runs.

2. A complete schedule of all of the symbols used with appropriate description of the requirements.

3. Riser diagram showing the physical relationship of the service, feeder and major power runs, units substations, isolated power transformers, switchboards, panel boards, pull boxes, terminal cabinets, and other systems and equipment.

4. Where necessary for clearness, a single line diagram showing the electrical relationship of the component items and sections of the wiring system.

5. Where necessary to provide adequate information elevations, sections and details of equipment and special installations and details of special lighting fixtures and devices.

6. Sections of the building or elevation of the structure showing floor-to-floor, outlet, and equipment heights, relation to the established grade, general type of building construction, etc. Where practical, suspended ceiling heights indicated by figure dimensions on either the electrical floor plan layout drawings or on the electrical building section or elevation drawings.

7. Where necessary to provide adequate information, plot plan to scale, showing the relation of the building or structure to other buildings or structures, service poles, service manholes, exterior area lighting, exterior wiring runs, etc.

8. In the case of exterior wiring systems for street and highway lighting, area drawings showing the complete system.

9. Any changes to the electrical layout should be clearly identified on the drawings when such changes are made after the original drawings have been completed and identified on the drawing by a revision symbol.

0.2 Explanation Supplementing the Schedule of Symbols

A. General

1. Type of Wiring Method or Material Requirement: When the general wiring method and material requirements for the entire installation are described in the specifications or specification notations on drawings, no special notation need be made in relation to symbols on the drawing layout, e.g., if an entire installation is required by the specifications and general reference on the drawings to be explosion-proof, the outlet symbols don't need to have special identification.

When certain different wiring methods or special materials will be required in different areas of the building or for certain sections of the wiring system or certain outlets, such requirements should be clearly identified on the drawing layout by special identification of outlet symbols rather than only by reference in the specifications.

2. Special Identification of Outlets: Weather-proof, vapor-tight, water-tight, rain-tight, dust-tight, explosion-proof, grounded or recessed outlets, or other special identification may be indicated by the use of uppercase letter abbreviations at the standard outlet symbol. See the examples in Table 3-1

Table 3-1   Abbreviations for Special Identification of Outlets

Weather-proof	WP
Vapor-tight	VT
Water-tight	WT

(continued)

Table 3-1   *(continued)*

Rain-tight	RT
Dust-tight	DT
Explosion-proof	EP
Grounded	G
Recessed	R

The grade, rating, and function of wiring devices used at special outlets should be indicated by abbreviated notation at the outlet location.

When the standard Special Purpose Outlet symbol is used to denote the location of special equipment or outlets or points of connection for such equipment, the specific usage will be identified by the use of a subscript numeral or letter alongside the symbol. The usage indicated by different subscripts will be noted on the drawing schedule of symbols.

B. Lighting Outlets

1. Identification of Type of Installation: A major variation in the type of outlet box, outlet supporting means, wiring system arrangement, and outlet connection and need of special items such as plaster rings or roughing-in cans often depend upon whether a lighting fixture is to be recessed or surface-mounted. A means of readily differentiating between such situations on drawings has been deemed to be necessary. In the case of a recessed-fixture installation, the standard adopted consists of a capital letter R drawn within the outlet symbol.

2. Fixture Identification: Lighting fixtures are identified as to type and size by the use of an uppercase letter, placed alongside each outlet symbol, together with a notation of the lamp size and number of lamps per fixture unit when two or more lamps per unit are required. A description of the fixture identified by the letter will be given either in the drawing schedule of symbols, separate fixture schedule on the drawing, or in the electrical specifications.

3. Switching Outlets: When different lighting outlets within a given local area are to be controlled by separately located wall switches, the related switching will be indicated by the use of lowercase letters at the lighting and switch outlet locations.

C. Signaling Systems

1. Basic Symbols: Each different basic category of signaling system shall be represented by a distinguishing Basic Symbol. Every item of equipment or outlet comparison in that category of system shall be identified by that basic symbol.

2. Identification of Individual Items: Different types of individual items of equipment or outlets indicated by a basic system symbol will be further identified by a numeral placed within the open system basic symbol. All such individual symbols used on the drawings shall be included on the drawing schedule of symbols.

3. Use of Symbols: Only the basic signaling system outlet symbols are included in this standard. The system or schedule of numbers referred to in (2) above will be developed by the designer.

4. Residential Symbols: Signaling system symbols for use in identifying certain specific standardized residential-type signal system items on residential drawings are included in this standard. The reason for this specific group of symbols is that a descriptive symbol list such as is necessary for the above group of basic system symbols is often not included on residential drawings.

D. Power Equipment

1. Rotating Equipment: At motor and generator locations, note on the drawing adjacent to the symbol the horsepower of each motor, or the capacity of each generator. When motors and generators of more than one type or system characteristic, e.g., voltage and phase, are required on a given installation, the specific types and system characteristics should be noted at the outlet symbol.

2. Switchboards, Power Control Centers, Unit Substations, and Transformer Vaults: The exact location of such equipment on the electrical layout floor plan drawing should be shown.

A detailed layout, including plan, elevation, and sectional views, should be shown when needed for clearness showing the relationship of such equipment to the building structure or other sections of the electric system.

A single line diagram, using American Standard Graphic Symbols for Electrical Diagrams—Y32.2, should be included to show the electrical relationship of the components of the equipment to each other and to the other sections of the electric system.

   E. Symbols Not Included in This Standard

   1. Certain electrical symbols that are commonly used in making electrical system layouts on drawings are not included as part of this standard for the reason that they have previously been included in American Standard Graphic Symbols for Electrical Diagrams, W32.2.

ASA policy requires that the same symbol not be included in two or more standards. The reason for this is that if the same symbol was included in two or more standards, when a symbol included in one standard was revised, it might not be so revised in the other standard at the same time, leading to confusion as to which was the proper symbol to use.

   2. Symbols falling into the above category include, but are not limited to, those shown in Table 3-2. The reference numbers are the American Standard Y32.2 item numbers.

Table 3-2   Symbols Included in Two or More Different Standards

(MOT)	46.3 Electric motor
(GEN)	46.2 Electric generator
(transformer symbol)	86.1 Power transformer
(pothead symbol)	82.1 Pothead (cable termination)
(WH)	48 Electric watt-hour meter
(CB)	12.2 Circuit element, e.g., circuit breaker
(circuit breaker symbol)	11.1 Circuit breaker

Table 3-2   *(continued)*

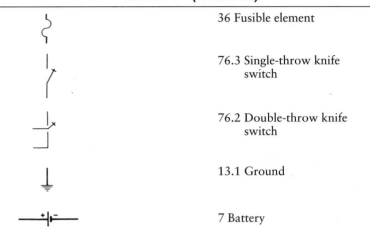

	36 Fusible element
	76.3 Single-throw knife switch
	76.2 Double-throw knife switch
	13.1 Ground
	7 Battery

List of Symbols

1.0 Lighting Outlets
See Table 3-3 for a list of lighting outlet symbols.

Table 3-3   Lighting Outlet Symbols

Ceiling	Wall	
◯	─◯	1.1 Surface or pendant incandescent mercury vapor or similar lamp fixture
Ⓡ	─Ⓡ	1.2 Recessed incandescent mercury vapor or similar lamp fixture
	[0]	1.3 Surface or pendant individual fluorescent fixture
	[OR]	1.4 Recessed individual fluorescent fixture
[0 | |]		1.5 Surface or pendant continuous-row fluorescent fixture

(continued)

Table 3-3 (continued)

Ceiling	Wall	
OR ⬜⬜⬜		1.6 Recessed continuous-row fluorescent fixture*
⊢—+—+—⊣		1.7 Bare-lamp fluorescent strip**
Ⓧ	—Ⓧ	1.8 Surface or pendant exit light
ⓍⓇ	—ⓍⓇ	1.9 Recessed exit light
Ⓑ	—Ⓑ	1.10 Blanked outlet
Ⓙ	—Ⓙ	1.11 Junction box
Ⓛ	—Ⓛ	1.12 Outlet controlled by low-voltage switching when relay is installed in outlet box

*In the case of combination continuous-row fluorescent and incandescent spotlights, use combinations of the above standard symbols.

**In the case of continuous-row, bare-lamp fluorescent strip above an area-wide diffusing means, show each fixture run, using the standard symbol; indicate area of diffusing means and type by light shading and/or drawing notation.

2.0 Receptacle Outlets

Where all or a majority of receptacles in an installation are to be of the grounding type, the uppercase-letter abbreviated notation may be omitted and the types of receptacles required noted in the drawing list of symbols and/or in the specifications. When this is done, any nongrounding receptacles may be so identified by notation at the outlet location.

Where weather-proof, explosion-proof, or other specific types of devices are required, use the type of uppercase subscript letters referred to under Section 0.2 item A-2 of this standard. For example, weather-proof single or duplex receptacles would have the uppercase subscript letters noted alongside of the symbol (see Table 3-4).

Table 3-4   Symbols for Receptacle Outlets

Ungrounded	Grounding	
		2.1 Single receptacle outlet.
		2.2 Duplex receptacle outlet.
		2.3 Triplex receptacle outlet.
		2.4 Quadruplex receptacle outlet.
		2.5 Duplex receptacle outlet—split wired.
		2.6 Triplex receptacle outlet—split wired.
		2.7 Single special-purpose receptacle outlet.*
		2.8 Duplex special-purpose receptacle outlet.*
		2.9 Range outlet.
		2.10 Special-purpose connection or provision for connection. Use subscript letters to indicate function (DW—dishwasher; CD—clothes dryer, etc.)
		2.11 Multi-outlet assembly. (Extend arrows to limit of installation. Use appropriate symbol to indicate type of outlet. Also indicate spacing of outlets as x inches.)
		2.12 Clock hanger receptacle.
		2.13 Fan hanger receptacle.

(continued)

Table 3-4 *(continued)*

Ungrounded	Grounding	

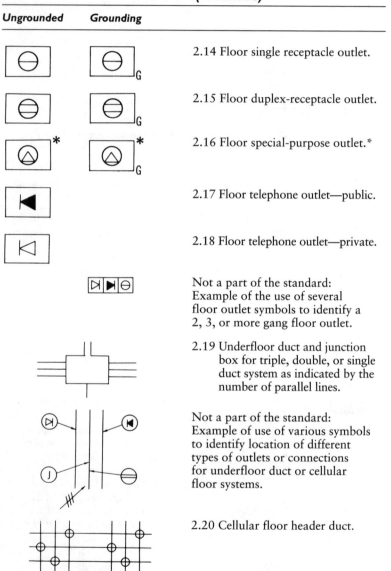

2.14 Floor single receptacle outlet.

2.15 Floor duplex-receptacle outlet.

2.16 Floor special-purpose outlet.*

2.17 Floor telephone outlet—public.

2.18 Floor telephone outlet—private.

Not a part of the standard:
Example of the use of several
floor outlet symbols to identify a
2, 3, or more gang floor outlet.

2.19 Underfloor duct and junction
box for triple, double, or single
duct system as indicated by the
number of parallel lines.

Not a part of the standard:
Example of use of various symbols
to identify location of different
types of outlets or connections
for underfloor duct or cellular
floor systems.

2.20 Cellular floor header duct.

*Use numeral or letter either within the symbol or as a subscript alongside the symbol in the drawing list of
symbols to indicate type of receptacle or usage.

3.0 Switch Outlets

See Table 3-5.

Table 3-5   Symbols for Switch Outlets

S	3.1 Single-pole switch
S_2	3.2 Double-pole switch
S_3	3.3 Three-way switch
S_4	3.4 Four-way switch
S_K	3.5 Key-operated switch
S_{LM}	3.6 Switch and pilot lamp
—⊖S	3.7 Switch for low-voltage switching system
═⊖S	3.8 Master switch for low-voltage switching system
S_D	3.9 Switch and single receptacle
S_T	3.10 Switch and double receptacle
S_{CB}	3.11 Door switch
	3.12 Time switch
S_{MC}	3.13 Circuit breaker switch
	3.14 Momentary contact switch or pushbutton for other than signaling system
Ⓢ	3.15 Ceiling pull switch

4.0 Signaling System Outlets: Institutional, Commercial, and Industrial Occupancies

See Table 3-6

Table 3-6   Symbols for Institutional, Commercial, and Industrial Occupancies

Basic Symbol	Examples of Individual Item Identification (Not a Part of the Standard)
—+○	4.1 Nurse Call System Devices (Any Type)

(continued)

Table 3-6 *(continued)*

Basic Symbol	*Examples of Individual Item Identification (Not a Part of the Standard)*	
	—+(1)	Nurses' annunciator (can add a number after it as —+(1) 24 to indicate number of lamps).
	—+(2)	Call station, single cord, pilot light.
	—+(3)	Call station, double cord, microphone-speaker.
	—+(4)	Corridor dome light, 1 lamp.
	—+(5)	Transformer.
	—+(6)	Any other item on same system—use numbers as required.
—+⬦		**4.2 Paging System Devices (Any Type)**
	—+⟨1⟩	Keyboard.
	—+⟨2⟩	Flush annunciator.
	—+⟨3⟩	2-Face annunciator.
	—+⟨4⟩	Any other item on same system—use numbers as required.

Table 3-6   *(continued)*

Basic Symbol	Examples of Individual Item Identification (Not a Part of the Standard)	
⊣▢		**4.3 Fire Alarm System Devices (Any Type) Including Smoke and Sprinkler Alarm Devices**
	⊣[1]	Control panel.
	⊣[2]	Station.
	⊣[3]	10" gong
	⊣[4]	Pre-signal chime.
	⊣[5]	Any other item on same system—use numbers as required.
⊣◇		**4.4 Staff Register System Devices (Any Type)**
	⊣◇1	Phone operators' register.
	⊣◇2	Entrance register—flush.
	⊣◇3	Staff room register.
	⊣◇4	Transformer.
	⊣◇5	Any other item on same system—use numbers as required.

(continued)

Table 3-6 *(continued)*

Basic Symbol	Examples of Individual Item Identification (Not a Part of the Standard)	
		4.5 Electric Clock System Devices (Any Type)
	①>	Master clock.
	②>	12" secondary—flush.
	③>	12" double dial—wall mounted.
	④>	18" skeleton dial.
	⑤>	Any other item on same system—use numbers as required.
		4.6 Public Telephone System Devices
	◁1	Switchboard.
	◁2	Desk phone.
	◁3	Any other item on same system—use numbers as required.
		4.7 Private Telephone System Devices (Any Type)
	◀1	Switchboard.
	◀2	Wall phone.

Table 3-6   *(continued)*

Basic Symbol	Examples of Individual Item Identification (Not a Part of the Standard)	
	$+$◀3	Any other item on same system—use numbers as required.
$+$⌂		**4.8 Watchman System Devices (Any Type)**
	$+$⌂1	Central station.
	$+$⌂2	Key station.
	$+$⌂3	Any other item on same system—use numbers as required.
$+$◁		**4.9 Sound System**
	$+$◁1	Amplifier.
	$+$◁2	Microphone.
	$+$◁3	Interior speaker.
	$+$◁4	Exterior speaker.
	$+$◁5	Any other item on same system—use numbers as required.
$+$◉		**4.10 Other Signal System Devices**
	$+$◉1	Buzzer.
	$+$◉2	Bell.

(continued)

Table 3-6   (continued)

Basic Symbol	Examples of Individual Item Identification (Not a Part of the Standard)	
	+–③	Pushbutton.
	+–④	Annunciator.
	+–⑤	Any other item on same system—use numbers as required.

5.0 Signaling System Outlets: Residential Occupancies

Table 3-7 presents signaling system symbols for use in identifying standardized residential-type signal system items on residential drawings where a descriptive symbol list is not included on the drawing. When other signal system items are to be identified, use the basic symbols presented in Table 3-7 for such items together with a descriptive symbol list.

Table 3-7   Signaling System Symbols for Residential Occupancies

▪	5.1 Pushbutton
◺	5.2 Buzzer
⊲	5.3 Bell
⊲/	5.4 Combination bell-buzzer
CH	5.5 Chime
◇—	5.6 Annunciator
D	5.7 Electric door opener

Table 3-7 *(continued)*

M	5.8 Maid's signal plug
☐	5.9 Interconnection box
BT	5.10 Bell-ringing transformer
▶│	5.11 Outside telephone
▷│	5.12 Interconnecting telephone
R	5.13 Radio outlet
TV	5.14 Television outlet

6.0 Panelboards, Switchboards, and Related Equipment
See Table 3-8.

Table 3-8   Symbols for Panelboards, Switchboards, and Related Equipment

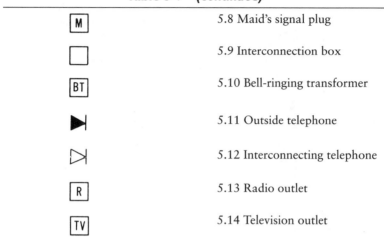

	6.1 Flush-mounted panelboard and cabinet.*
	6.2 Surface-mounted panelboard and cabinet.*
	6.3 Switchboard, power control center, unit substations*— should be drawn to scale.
TC	6.4 Flush-mounted terminal cabinet.* (In small-scale drawings the TC may be indicated alongside the symbol.)

(continued)

Table 3-8 (continued)

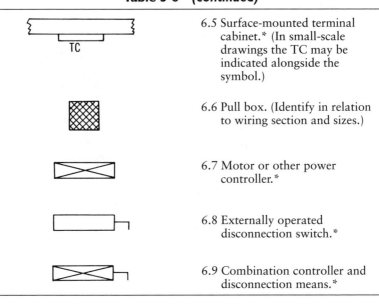

	6.5 Surface-mounted terminal cabinet.* (In small-scale drawings the TC may be indicated alongside the symbol.)
	6.6 Pull box. (Identify in relation to wiring section and sizes.)
	6.7 Motor or other power controller.*
	6.8 Externally operated disconnection switch.*
	6.9 Combination controller and disconnection means.*

Identify by notation or schedule.

7.0 Bus Ducts and Wireways
See Table 3-9.

Table 3-9   Symbols for Bus Ducts and Wireways

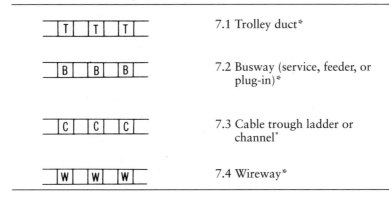

	7.1 Trolley duct*
	7.2 Busway (service, feeder, or plug-in)*
	7.3 Cable trough ladder or channel*
	7.4 Wireway*

8.0 Remote Control Stations for Motors or Other Equipment*
See Table 3-10.

Table 3-10   Symbols for Remote Control Stations for Motors of Other Equipment

	8.1 Pushbutton station
F	8.2 Float switch—mechanical
L	8.3 Limit switch—mechanical
P	8.4 Pneumatic switch—mechanical
	8.5 Electric eye—beam source
	8.6 Electric eye—relay
T	8.7 Thermostat

*Identify by notation or schedule.

9.0 Circuiting
Wiring method identification by notation on drawing or in specification (Table 3-11).

Table 3-11   Symbols for Circuiting

	9.1 Wiring concealed in ceiling or wall.
	9.2 Wiring concealed in floor.
	9.3 Wiring exposed.

(continued)

Table 3-11   (continued)

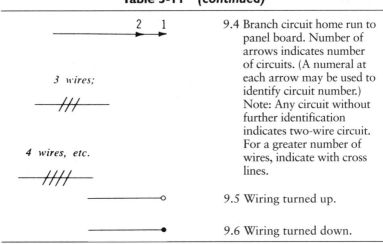

	9.4 Branch circuit home run to panel board. Number of arrows indicates number of circuits. (A numeral at each arrow may be used to identify circuit number.) Note: Any circuit without further identification indicates two-wire circuit. For a greater number of wires, indicate with cross lines.
	9.5 Wiring turned up.
	9.6 Wiring turned down.

Note: Use heavy-weight line to identify service and feeders. Indicate empty conduit by notation CO (conduit only). Unless indicated otherwise, the wire size of the circuit is the minimum size required by the specification. Identify different functions of wiring system, e.g., signaling system, by notation or other means.

10.0 Electric Distribution or Lighting System, Underground
See Table 3-12.

Table 3-12   Symbols for Underground Electric Distribution or Lighting System

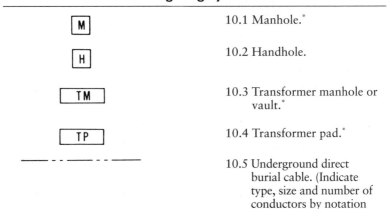

	10.1 Manhole.*
	10.2 Handhole.
	10.3 Transformer manhole or vault.*
	10.4 Transformer pad.*
	10.5 Underground direct burial cable. (Indicate type, size and number of conductors by notation or schedule.)

Table 3-12   (continued)

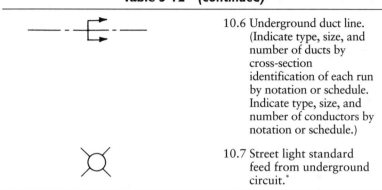

	10.6 Underground duct line. (Indicate type, size, and number of ducts by cross-section identification of each run by notation or schedule. Indicate type, size, and number of conductors by notation or schedule.)
	10.7 Street light standard feed from underground circuit.*

*Identify by notation or schedule.

11.0 Electric Distribution or Lighting System Aerial
See Table 3-13.

Table 3-13   Symbols for Electric Distribution or Lighting System Aerial

◯	11.1 Pole*
⊘⊢	11.2 Street light and bracket*
△	11.3 Transformer*
———	11.4 Primary circuit*
- - - - -	11.5 Secondary circuit*
——→)	11.6 Down guy
——●—	11.7 Head guy
——○—)	11.8 Sidewalk guy
⊄——	11.9 Service weather head*

*Identify by notation or schedule.

Arrester, Lightning Arrester (Electric Surge, etc.) Gap
See Table 3-14.

Table 3-14 Symbols for Arrester, Lightning Arrester Gap

	4.1 General.
	4.2 Carbon block. Block, telephone protector. The sides of the rectangle are to be approximately in the ratio of 1 to 2 and the space between rectangles shall be approximately equal to the width of a rectangle.
	4.3 Electrolytic or aluminum cell. This symbol is not composed of arrowheads.
	4.4 Horn gap.
	4.5 Protective gap. These triangles shan't be filled.
	4.6 Sphere gap.
	4.7 Valve or film element.
	4.8 Multigap, general.
	4.9 Application: gap plus valve plus ground, 2 pole.

*Identify by notation or schedule.

Battery
See Table 3-15.

Table 3-15   Symbols for Batteries

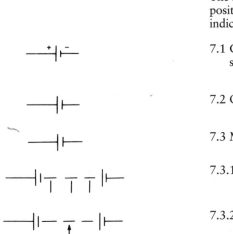

	The long line is always positive, but polarity may be indicated in addition. Example:
	7.1 Generalized direct-current source
	7.2 One-cell
	7.3 Multicell
	7.3.1 Multicell battery with 3 taps
	7.3.2 Multicell battery with adjustable tap

Circuit Breakers
See Table 3-16.

Table 3-16   Symbols for Circuit Breakers

	If it is desired to show the condition causing the breaker to trip, the relay-protective-function symbols in item 66.6 may be used alongside the break symbol.
	11.1 General.
	11.2 Air circuit breaker, if distinction is needed; for alternating-current breakers rated at 1500 volts or less and for all direct-current circuit breakers.

(continued)

Table 3-16 *(continued)*

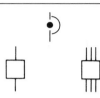

11.2.1 Network protector.

See note 11.3A

11.3 Circuit breaker, other than covered by item 11.2. The symbol in the right column is for a 3-pole breaker.

See note 11.3A

11.3.1 On a connection or wiring diagram, a 3-pole single-throw circuit breaker (with terminals shown) may be drawn as shown.

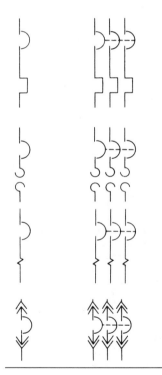

11.4 Applications.

11.4.1 3-pole circuit breaker with thermal overload device in all 3 poles

11.4.2 3-pole circuit breaker with magnetic overload device in all 3 poles.

11.4.3 3-pole circuit breaker, draw-out type.

Note 11.3A—On a power diagram, the symbol may be used without other identification. On a composite drawing where confusion with the general circuit element symbol (item 12) may result, add the identifying letters CB inside or adjacent to the square.

Circuit Return

See Table 3-17.

Table 3-17   *Symbols for Circuit Return*

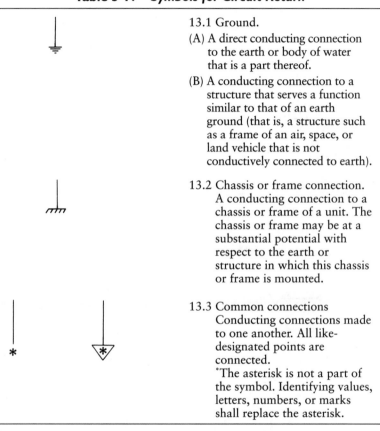

	13.1 Ground.
	(A) A direct conducting connection to the earth or body of water that is a part thereof.
	(B) A conducting connection to a structure that serves a function similar to that of an earth ground (that is, a structure such as a frame of an air, space, or land vehicle that is not conductively connected to earth).
	13.2 Chassis or frame connection. A conducting connection to a chassis or frame of a unit. The chassis or frame may be at a substantial potential with respect to the earth or structure in which this chassis or frame is mounted.
	13.3 Common connections Conducting connections made to one another. All like-designated points are connected. *The asterisk is not a part of the symbol. Identifying values, letters, numbers, or marks shall replace the asterisk.

Coil, Magnetic Blowout*

---

*The broken line indicates where line connection to a symbol is made and is not a part of the symbol.

Contact, Electrical
See Table 3-18.

Table 3-18   Symbols for Electrical Contact

	23.1 Fixed contact.
	23.1.1 Fixed contact for jack, key, relay, etc.
	23.1.2 Fixed contact for switch.
	23.1.3 Fixed contact for momentary switch. See SWITCH (item 76.8 and 76.10).
	23.1.4 Sleeve.
	23.2 Moving contact.
	23.2.1 Adjustable or sliding contact for resistor, inductor, etc.
	23.2.2 Locking.
	23.2.3 Segment; bridging contact.
	23.2.4 Nonlocking. See SWITCH (items 76.12.3 and 76.12.4).
	23.2.5 Vibrator reed.
	23.2.6 Vibrator split reed.
	23.2.7 Rotating contact (slip ring) and brush.
	23.3 Basic contact assemblies.

Table 3-18   Symbols for Electrical Contact

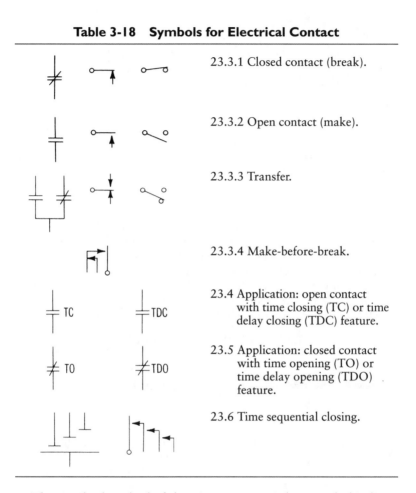

	23.3.1 Closed contact (break).
	23.3.2 Open contact (make).
	23.3.3 Transfer.
	23.3.4 Make-before-break.
	23.4 Application: open contact with time closing (TC) or time delay closing (TDC) feature.
	23.5 Application: closed contact with time opening (TO) or time delay opening (TDO) feature.
	23.6 Time sequential closing.

The standard method of showing a contact is by a symbol indicating the circuit condition it produces when the actuating device is in the deenergized or nonoperated position. The actuating device may be of a mechanical, electrical, or other nature, and a clarifying note may be necessary with the symbol to explain the proper point at which the contact functions; for example, the point where a contact closes or opens as a function of changing pressure, level, flow, voltage, current, etc. In cases where it is desirable to show contacts in the energized or operated condition and where confusion may result, a clarifying note shall be added to the drawing. Auxiliary switches or contacts for circuit breakers, etc., may be designated as follows:

(a) Closed when device is in energized or operated position

(b) Closed when device is in deenergized or nonoperated position

(aa) Closed when operating mechanism of main device is in energized or operated position

(bb) Closed when operating mechanism of main device is in deenergized or nonoperated position

    In the parallel-line contact symbols showing the length of the parallel lines shall be approximately 1¼ times the width of the gap (except for item 23.6).

Contactor

See also RELAY (item 66).

    Fundamental symbols for contacts, coils, mechanical connections, etc., are the basis of contactor symbols and should be used to represent contactors on complete diagrams (Table 3-19). Complete diagrams of contactors consist of combinations of fundamental symbols for control coils, mechanical connections, etc., in such configurations as to represent the actual device.

    Mechanical interlocking should be indicated by notes.

Table 3-19   Contactor Symbols

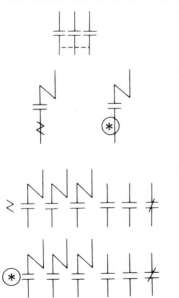

24.1 Manually operated 3-pole contactor.

24.2 Electrically operated 1-pole contactor with series blowout coil.
*See Note 24.2A.

24.3 Electrically operated 3-pole contactor with series blowout coils; 2 open and 1 closed auxiliary contacts (shown smaller than the main contacts).

Table 3-19   *(continued)*

		24.4 Electrically operated 1-pole contactor with shunt blowout coil.

Note 24.2A—The asterisk is not a part of the symbol. Always replace the asterisk by a device designation.

Machine, Rotating
See Table 3-20.

Table 3-20   Symbols for Rotating Machines

	46.1 Basic.
	46.2 Generator (general).
	46.3 Motor (general).
	46.4 Motor, multispeed. Use Basic Motor Symbol and note speeds.
	46.5 Rotating armature with commutator and brushes.*
	46.6 Field, generator, or motor. Either symbol of item 42.1 may be used in the following items.
	46.6.1 Compensating or commutating.
	46.6.2 Series.
	46.6.3 Shunt or separately excited.
	46.6.4 Magnet, permanent. See item 47.

(continued)

Table 3-20 *(continued)*

46.7 Winding symbols.
Motor and generator winding symbols may be shown in the basic circle using the following representation.

46.7.1 1-phase.

46.7.2 2-phase.

46.7.3 3-phase wye (ungrounded).

46.7.4 3-phase wye (grounded).

46.7.5 3-phase delta.

46.7.6 6-phase diametrical.

46.7.7 6-phase double-delta.

46.8 Direct-current machines; applications.

46.8.1 Separately excited direct-current generator or motor.[*]

46.8.2 Separately excited direct-current generator or motor; with commutating or compensating field winding or both.[*]

46.8.3 Compositely excited direct-current generator or motor; with commutating or compensating field winding or both.[*]

Table 3-20 *(continued)*

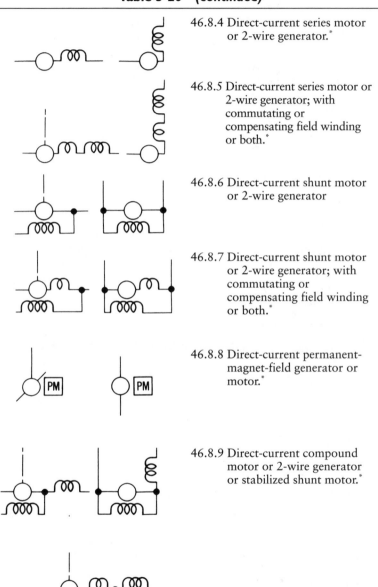

46.8.4 Direct-current series motor or 2-wire generator.[*]

46.8.5 Direct-current series motor or 2-wire generator; with commutating or compensating field winding or both.[*]

46.8.6 Direct-current shunt motor or 2-wire generator

46.8.7 Direct-current shunt motor or 2-wire generator; with commutating or compensating field winding or both.[*]

46.8.8 Direct-current permanent-magnet-field generator or motor.[*]

46.8.9 Direct-current compound motor or 2-wire generator or stabilized shunt motor.[*]

(continued)

Table 3-20 *(continued)*

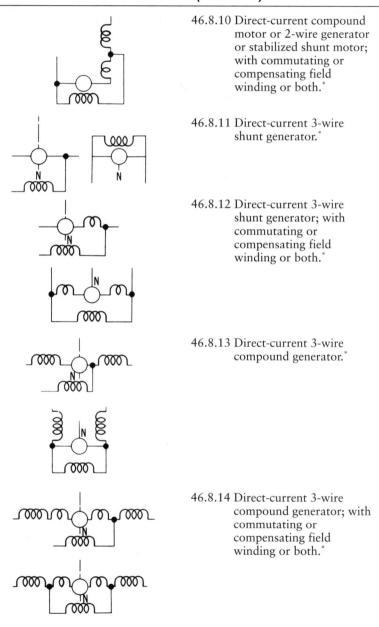

46.8.10 Direct-current compound motor or 2-wire generator or stabilized shunt motor; with commutating or compensating field winding or both.*

46.8.11 Direct-current 3-wire shunt generator.*

46.8.12 Direct-current 3-wire shunt generator; with commutating or compensating field winding or both.*

46.8.13 Direct-current 3-wire compound generator.*

46.8.14 Direct-current 3-wire compound generator; with commutating or compensating field winding or both.*

Table 3-20 (continued)

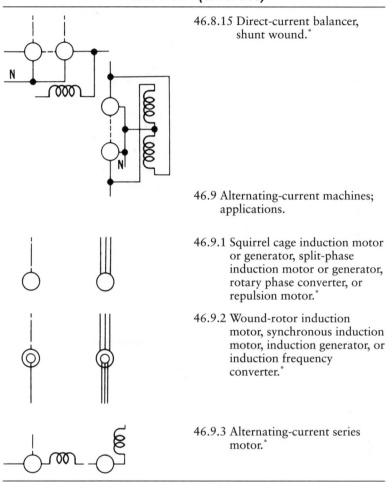

46.8.15 Direct-current balancer, shunt wound.*

46.9 Alternating-current machines; applications.

46.9.1 Squirrel cage induction motor or generator, split-phase induction motor or generator, rotary phase converter, or repulsion motor.*

46.9.2 Wound-rotor induction motor, synchronous induction motor, induction generator, or induction frequency converter.*

46.9.3 Alternating-current series motor.*

*The broken line indicates where line connection to a symbol is made and is not a part of the symbol.

Meter Instrument

Note 48A—The asterisk is not a part of the symbol. Always replace the asterisk by one of the following letter combinations (Table 3-21), depending on the function of the meter or instrument, unless some other identification is provided in the circle and explained on the diagram.

Table 3-21   Symbols for Meter Instruments

A	Ammeter IEC
AH	Ampere-hour
CMA	Contact-making (or -breaking) ammeter
CMC	Contact-making (or -breaking) clock
CMV	Contact-making (or -breaking) voltmeter
CRO	Oscilloscope or cathode-ray oscillograph
DB	DB (decibel) meter
DBM	DBM (decibels referred to 1 milliwatt) meter
DM	Demand meter
DTR	Demand-totalizing relay
F	Frequency meter
G	Galvanometer
GD	Ground detector
I	Indicating
INT	Integrating
μA or UA	Microammeter
MA	Milliammeter
NM	Noise meter
OHM	Ohmmeter
OP	Oil pressure
OSCG	Oscillograph string
PH	Phasemeter
PI	Position indicator
PF	Power factor
RD	Recording demand meter
REC	Recording
RF	Reaction factor
SY	Synchroscope
TLM	Telemeter
T	Temperature meter
THC	Thermal converter
TT	Total time
V	Voltmeter

Table 3-21 (continued)

VA	Volt-ammeter
VAR	Varmeter
VARH	Varhour meter
VI	Volume indicator; meter, audio level
VU	Standard volume indicator; meter, audio level
W	Wattmeter
WH	Watthour meter

Path, Transmission, Conductor, Cable Wiring
See Table 3-22.

Table 3-22 Symbols for Path, Transmission, Conductor, and Cable Wiring

58.1 Guided path, general.
A single line represents the entire group of conductors or the transmission path needed to guide the power or the signal. For coaxial and waveguide work, the recognition symbol is used at the beginning and end of each kind of transmission path and at intermediate points as needed for clarity. In waveguide work, mode may be indicated.

58.2 Conductive path or conductor; wire.

58.2.1 Two conductors or conductive paths of wires.

58.2.2 Three conductors or conductive paths of wires.

58.2.3 "n" conductors or conductive paths of wires.
"n" conductors.

(continued)

Table 3-22 *(continued)*

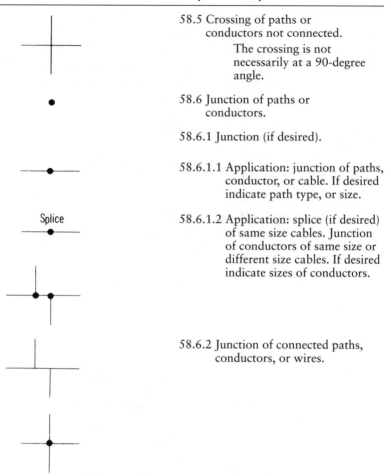

	58.5 Crossing of paths or conductors not connected. The crossing is not necessarily at a 90-degree angle.
	58.6 Junction of paths or conductors.
	58.6.1 Junction (if desired).
	58.6.1.1 Application: junction of paths, conductor, or cable. If desired indicate path type, or size.
Splice	58.6.1.2 Application: splice (if desired) of same size cables. Junction of conductors of same size or different size cables. If desired indicate sizes of conductors.
	58.6.2 Junction of connected paths, conductors, or wires.

Polarity Symbol
See Table 3-23.

Table 3-23   Polarity Symbols

+	63.1 Positive
−	63.2 Negative

Switch

See Table 3-24.

Fundamental symbols for contacts, mechanical connections, etc., may be used for switch symbols.

The standard method of showing switches is in a position with no operating force applied. For switches that may be in any one of two or more positions with no operating force applied and for switches actuated by some mechanical device (as in air-pressure, liquid-level, rate-of-flow, etc., switches), a clarifying note may be necessary to explain the point at which the switch functions.

When the basic switch symbols in items 76.1 through 76.4 are shown on a diagram in the closed position, terminals must be added for clarity.

Table 3-24   Switch Symbols

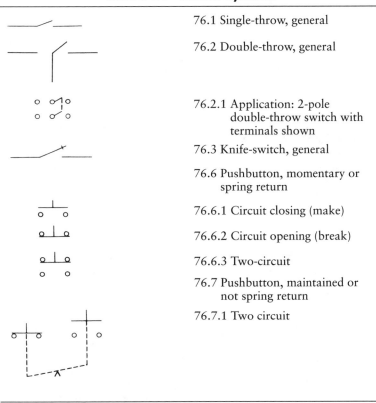

	76.1 Single-throw, general
	76.2 Double-throw, general
	76.2.1 Application: 2-pole double-throw switch with terminals shown
	76.3 Knife-switch, general
	76.6 Pushbutton, momentary or spring return
	76.6.1 Circuit closing (make)
	76.6.2 Circuit opening (break)
	76.6.3 Two-circuit
	76.7 Pushbutton, maintained or not spring return
	76.7.1 Two circuit

Transformer
See Table 3-25.

Table 3-25   Transformer Symbols

	86.1 General.
	Either winding symbol may be used. In the following items, the left symbol is used. Additional windings may be shown or indicated by a note. For power transformers use polarity marking H_1, X_1, etc., from American Standard C6.1-1956.
	For polarity markings on current and potential transformers, see items 86.16.1 and 86.17.1.
	In coaxial and waveguide circuits, this symbol will represent a taper or step transformer without mode change.
	86.1.1 Application: transformer with direct-current connections and mode suppression between two rectangular waveguides.
	86.2 If it is desired especially to distinguish a magnetic-core transformer.
	86.2.1 Application: shielded transformer with magnetic core shown.
	86.2.2 Application: transformer with magnetic core shown and with a shield between windings. The shield is shown connected to the frame.

Table 3-25 *(continued)*

	86.6 With taps, 1-phase.
	86.7 Autotransformer, 1-phase.
	86.7.1 Adjustable.
	86.13 1-phase 2-winding transformer.
	86.13.1 3-phase bank of 1-phase 2-winding transformers. See American Standard C6.1-1965 for interconnections for complete symbol.
	86.14 Polyphase transformer.
	86.16 Current transformer(s).

(continued)

Table 3-25 *(continued)*

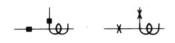

86.16.1 Current transformer with polarity marking. Instantaneous direction of current into one polarity mark corresponds to current out of the other polarity mark.

Symbol used shan't conflict with item 77 when used on same drawing.

86.16.2 Bushing-type current transformer.*

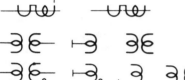

86.17 Potential transformer(s).

86.17.1 Potential transformer with polarity mark. Instantaneous direction of current into one polarity mark corresponds to current out of the other polarity mark.
Symbol used shan't conflict with item 77 when used on same drawing.

86.18 Outdoor metering device.

86.19 Transformer winding connection symbols.
For use adjacent to the symbols for the transformer windings.

86.19.1 2-phase 3-wire, ungrounded.

Table 3-25   *(continued)*

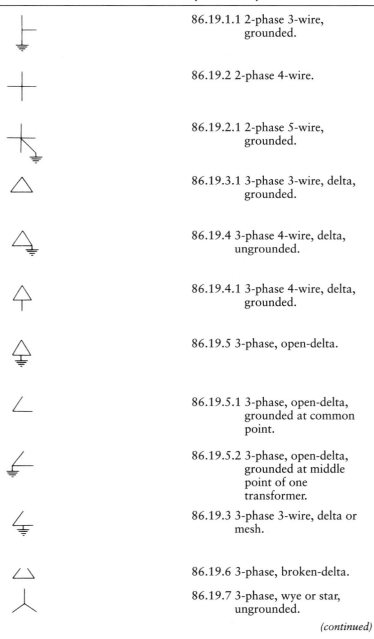

86.19.1.1 2-phase 3-wire, grounded.

86.19.2 2-phase 4-wire.

86.19.2.1 2-phase 5-wire, grounded.

86.19.3.1 3-phase 3-wire, delta, grounded.

86.19.4 3-phase 4-wire, delta, ungrounded.

86.19.4.1 3-phase 4-wire, delta, grounded.

86.19.5 3-phase, open-delta.

86.19.5.1 3-phase, open-delta, grounded at common point.

86.19.5.2 3-phase, open-delta, grounded at middle point of one transformer.

86.19.3 3-phase 3-wire, delta or mesh.

86.19.6 3-phase, broken-delta.

86.19.7 3-phase, wye or star, ungrounded.

(continued)

Table 3-25 (continued)

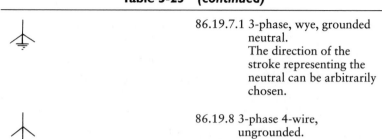

	86.19.7.1 3-phase, wye, grounded neutral. The direction of the stroke representing the neutral can be arbitrarily chosen.
	86.19.8 3-phase 4-wire, ungrounded.

*The broken line indicates where line connection to a symbol is made and is not a part of the symbol.

Chapter 4

Magnets and Magnetic Fields

In Asia Minor, and particularly in Magnesia, there is found a hard lead-colored mineral called *magnetite*. This is in reality iron oxide (Fe_3O_4), composed of three atoms of iron and four atoms of oxygen. In early times, people knew that this substance would attract small particles of iron. Later it was found that a piece of magnetite would always turn to point in one direction. Tradition states that Hoang-ti, a Chinese navigator, used a piece of magnetite floated on water to navigate a fleet of ships when out of sight of land. This was about 2400 B.C.

This substance was later found to exist in many places throughout the world and was called *lodestone* or *leading stone*.

In 1600 A.D. Dr. Gilbert conducted extensive research and published an account of his magnetic discoveries. In these he found that the attractive force appeared at two regions on the lodestone, which he designated as the *poles*. The region between the two poles becomes less magnetic and a point may be reached at which no magnetic forces exist. This may be illustrated by taking a magnetized bar and dipping it in iron filings (see Figure 4-1). The filings accumulate at the ends and not at the middle of the bar. This middle region on the bar was termed by Dr. Gilbert as the *equator*.

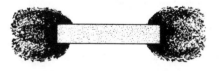

Figure 4-1 Magnetized iron bar and its attraction for iron filings.

In 1729, Savery discovered that hard steel retained magnetism better than soft iron.

Magnetic Poles

You may observe the effect of magnetism by means of a magnetic needle and a bar magnet. See Figure 4-2. You will observe that one end of the needle is attracted to one and only one end of the bar magnet and repelled by the other end of the bar magnet. It is sufficient for our purpose to characterize this difference in terms of poles, calling one end the *north pole* and the other end the *south pole*.

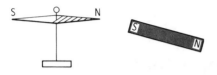

Figure 4-2 Magnetic needle and bar magnet.

Confusion comes out of this, as we will note that the south pole of the magnet attracts the north pole of the needle, while the north pole of the magnet repels the north pole of the needle. From this we deduce that like poles repel and unlike poles attract. This confusion comes from the fact that the north pole of a compass always points to the north magnetic pole of the earth. This tends to disrupt our attraction and repulsion observations. The French and Chinese call the north-pointing end of the compass the south pole. However it is customary in our country to call that pole of the magnet that points north if suspended the north pole, and the end that points to the south magnetic pole, the south pole.

We have just learned the first law of magnetism: Like poles repel and unlike poles attract.

Magnetic and Nonmagnetic Substances

A distinction must be made between magnets and magnetic substances. A magnet is a magnetic material in which polarity has been developed. Substances that are affected by magnets, that is, attracted by magnets, are magnetic substances. Some of these may retain magnetism and some others may not. Iron and steel are the most highly magnetic substances. Nickel is a magnetic substance to a degree.

All substances may be classified magnetically as follows:

1. Magnetic substance: those that are attracted by a magnet.
2. Diamagnetic substances: those that are feebly repelled by a magnet.
3. Nonmagnetic substances: those that are not affected by a magnet in any way.

The Earth as Magnet

The earth itself is a great magnet, with its magnetic poles nearly coinciding with the geographic poles. Thus we find that at Denver the compass points about 14° east of the true north-south line. At Savannah, the two directions coincide, and at New York the magnetic variation from the true north-south line is about 11° west of the true line. These variations change from year to year, and this information is readily available.

If a steel needle is magnetized and balanced by a thread, it is free to move in any direction. It is free to align itself vertically with the declination of the earth's magnetic field. The angle of dip increases the closer you get to the earth's magnetic poles. Figure 4-3 illustrates a dipping needle.

Figure 4-3   A dipping needle.

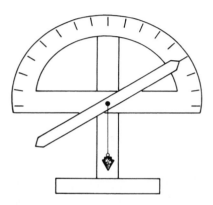

Magnetic Lines of Force

If an unmagnetized iron bar is approached by a pole of a magnet, the iron bar becomes magnetized, even though it is not touched by the magnet. This is called *magnetic induction* and if the amount of magnetism is comparatively high—that is, the magnetism induced into the iron bar—the bar is said to possess *high permeability.*

Figure 4-4 illustrates the lines of force that radiate from a bar magnet. Since the lines of magnetic force may neither be seen nor felt, a good method of proof that they exist is shown in Figure 4-5. A sheet of paper is placed over various magnets; the paper is then sprinkled with iron filings and tapped sharply. The filings align themselves with the magnetic lines of force.

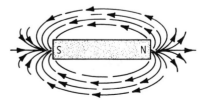

Figure 4-4   Direction of lines of magnetic force about a bar magnet.

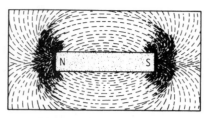

Figure 4-5   Use of iron filings to illustrate magnetic field.

(A) Magnetic field between unlike poles.

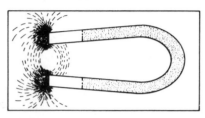

(B) Magnetic field around a horseshoe magnet.

There is no good insulator of magnetism. If, however, a hollow iron cylinder is placed in a magnetic field, the magnetic lines of force will follow a path through the iron cylinder and will be deflected away from the interior. See Figure 4-6.

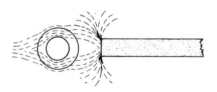

Figure 4-6   Deflection of magnetic lines of force.

Molecular Theory of Magnetism

If a bar magnet, such as shown in Figure 4-7, is broken into pieces, each piece will become a magnet within itself, with both north and south poles. The broken bar magnet will produce as many magnets as there are pieces into which it is broken.

(A) Original bar magnet.

N   S

(B) Broken bar magnet.

N   S   N   S   N   S

Figure 4-7   A bar magnet, when broken, produces as many magnets as there are pieces.

Joule found that an iron bar increases slightly in length when strongly magnetized and Bidwell found that still stronger magnetization causes the iron to subsequently shrink in length. If a bar is rapidly magnetized, first in one direction and then in the other, the bar will heat and a decided hum in the bar will be heard.

Experiments such as these have led to the theory that magnetism is an action affecting the arrangement of the molecules. Figure 4-8 shows the molecules of an unmagnetized bar of iron. It is theorized

Figure 4-8   Chaotic arrangement of molecules in an unmagnetized iron bar.

that when an iron bar is magnetized, its molecules are magnetized and line up. See Figure 4-9. High-carbon steel was the original material used in making permanent magnets.

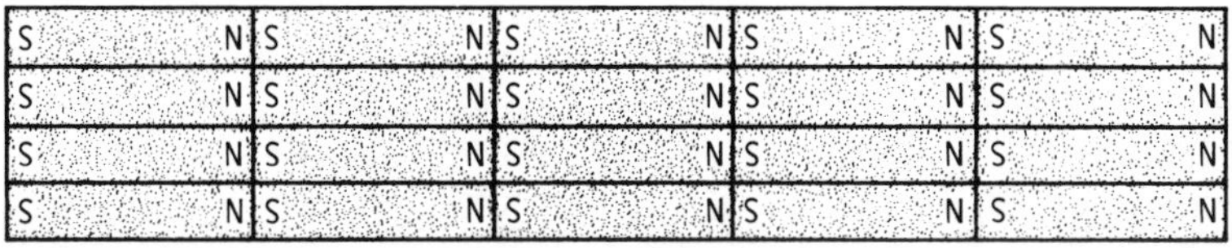

Figure 4-9   Arrangement of molecules in a magnet.

Magnets that are stronger and more permanent may be made from special alloys of steel containing tungsten, chromium, cobalt, aluminum, and nickel.

The following list illustrates some of the percentages of these alloys that are added to steel in the making of these stronger magnets:

Tungsten Steel		Chromium Steel		Cobalt Steel		Alnico Steel	
Tungsten	5–6%	Chromium	1–6%	Cobalt	3.5–	Aluminum	10–12%
Carbon	0.7%	Carbon	0.6–1.0%		8.0%	Nickel	17–28%
Manganese	0.3%	Manganese	0.2–0.6%	Chromium	3–9%	Cobalt	5–13%
Chromium	0.3%			Tungsten	1–9%		
				Carbon	0.9%		
				Manganese	0.3–		
					0.8%		

Strength of a Magnet

Soft iron will magnetize while under the influence of a magnetic field, but will retain very little, if any, magnetism when the magnetic field is removed.

We have also just seen that the alloy content of steel affects the magnetic strength.

The strength of a magnet may be determined by the magnetic force that it exhibits at a distance from other magnets. Thus, suppose there are two magnets acting upon a suspended needle. If at the same distance from the needle, the two magnets produce equal deflections, their strengths would be equal. In other words, the strength of a magnet may be defined as the amount of free magnetism at its pole.

Lifting Power of a Magnet

A distinction must be drawn between the *strength* of a magnet and its *lifting power.* The lifting power of a magnet depends upon the shape of its pole and the number of lines of force passing through its pole. A horseshoe magnet with both poles connected by a soft iron keeper will lift three to four times as much as with one pole alone. A bar magnet will lift more or less depending upon the shape of its poles. Suppose that there are three bar magnets of equal strength (Figure 4-10), and that there is the same amount of free magnetism at the poles of all three. If the end of bar A flares out, it

(A) Flared end pole.

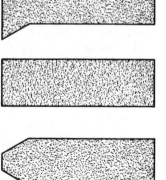

(B) Equal end pole.

(C) Chamfered end pole.

Figure 4-10 Illustration of different pole shapes that give different lifting powers.

will lift the least. If pole *B* is the same area as the bar, it will lift more. If the pole is chamfered off so that its face has a smaller cross section than the bar, as in *C*, it will lift the most. The reason for the various lifting powers mentioned above is found in the law that governs the lifting power of a magnet. This law states: "The lifting power is proportional to the square of the number of magnetic lines of force per unit of cross-section." Thus, if the pole is chamfered off until the area is reduced by one-half, and the original amount of magnetism is crowded into this reduced area, the flux density of the magnetism would be doubled. This would cause the lifting power for the reduced area to be quadrupled. Since the cross-section has been reduced by one-half, the lifting power is actually only doubled. Practically it would be impossible to concentrate the magnetism to such an extent as to bring about this result. Nevertheless, the lifting power may often be increased to some extent by diminishing the cross-section of the pole.

Questions

1. What is magnetite?
2. Complete the statement: Like poles_____each other.
3. Complete the statement: Unlike poles_____each other.
4. The north pole of a compass is attracted by which magnetic pole?
5. What is a magnetic substance?
6. What is a diamagnetic substance?
7. What is a nonmagnetic substance?
8. In your area, which way will a dipping needle dip?
9. Name an insulator of magnetism.
10. When you break a bar magnet, what happens to the magnetism?
11. Describe the molecular theory of magnetism.
12. What alloys are used in Alnico magnets?

Chapter 5
Ohm's Law

Georg Simon Ohm was born in 1789. His father taught him mathematics and the locksmith trade. He loved electricity and research work and published "The Galvanic Battery Treated Mathematically," which became a classic of science. In 1781, a law governing current in an electrical circuit was discovered by Cavendish, but this law was not publicly recognized until Ohm obtained results by experiment, for which he received from England in 1841, a gold medal for the "most conspicuous discovery in the domain of exact investigation."

The law that bears his name is Ohm's law. This law gives the relationship between voltage, current, and resistance in an electrical circuit. This law is accurate and absolute. When we come to alternating current, we will find that it still applies, with a few other factors being involved.

Statement of Ohm's Law

The fundamental statement of Ohm's law is as follows:

The current in amperes in any electrical circuit is numerically equal to the electromotive force (emf or voltage) in volts impressed upon that circuit, divided by the entire resistance of the circuit in ohms. The equation may be expressed

$$I = \frac{E}{R}$$

where

I = intensity of current in amperes

E = emf in volts

R = resistance in ohms

If any two quantities of this equation are known, the third quantity may be found by transposing, as

$$E = IR \quad R = E/I \quad \text{and} \quad I = E/R$$

Analogy of Ohm's Law

Hydraulic analogies may be used to illustrate electrical currents and the effect of friction (resistance). We all know from experience that water running through a hose encounters resistance. Thus with X pounds of pressure at the hydrant, we will get a considerable flow of water directly out of the open hydrant. If we use 50 ft of ¾-in hose, the flow will be cut down, and if we use 100 ft of ¾-in hose, the flow will be cut even more. The same will be true if we use 50-ft-long ½-in hose: We get less water with the same X pounds of pressure at the hydrant than we did with the ¾-in hose. The X pounds of pressure may be compared to volts. The quantity of water delivered may be compared to current (amperes), and the friction (resistance) of the various hoses may be compared to the resistance of the conductor in ohms.

Illustrations of Ohm's Law

Now, if we deliver 100 volts to the circuit in Figure 5-1A and the load resistance is 50 ohms, there will be a current of 2 amperes:

$$I = \frac{E}{R} = \frac{100\,V}{50\,\Omega} = 2\,A$$

In Figure 5-1B, we have the same voltage but the load resistance is 100 ohms, so we find

$$I = \frac{E}{R} = \frac{100\,V}{100\,\Omega} = 100\,A$$

Then, in Figure 5-1C, we have doubled the voltage to 200 volts and have 50 ohms of load resistance, so

$$I = \frac{E}{R} = \frac{200\,V}{50\,\Omega} = 4\,A$$

Here we see that doubling the voltage on the same load as in Figure 5-1A will double the current from 2 to 4 amperes.

Problem I
An incandescent lamp is connected on a 110-volt system. The resistance (R) of the heated filament is 275 ohms. What amount of current will the lamp draw? Answer:

$$I = \frac{E}{R} = \frac{110\,V}{275\,\Omega} = 0.4\,A$$

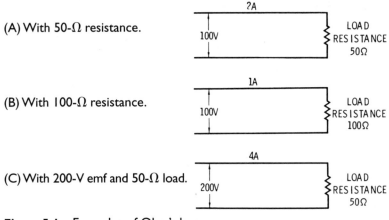

(A) With 50-Ω resistance.

(B) With 100-Ω resistance.

(C) With 200-V emf and 50-Ω load.

Figure 5-1   Examples of Ohm's law.

Problem 2

We have two electric heaters with resistances of 20 and 40 ohms, respectively. We also have 120 and 240 volts available. Calculate current amperes for the following combinations:

a. 20 ohms on 120 volts: $I = \dfrac{E}{R} = \dfrac{120\,V}{20\,\Omega} = 6\,A$

b. 40 ohms on 120 volts: $I = \dfrac{E}{R} = \dfrac{120\,V}{40\,\Omega} = 3\,A$

c. 20 ohms on 240 volts: $I = \dfrac{E}{R} = \dfrac{240\,V}{20\,\Omega} = 12\,A$

d. 40 ohms on 240 volts: $I = \dfrac{E}{R} = \dfrac{240\,V}{40\,\Omega} = 6\,A$

From the above problems, you may readily see that doubling the resistance on the same voltage halves the current; also, doubling the voltage with the same resistance doubles the current.

From this the deduction may be made that the current in any circuit will vary directly with the emf and also that the current in an electrical circuit varies inversely with the resistance. Also, if the current in an electrical circuit is to be maintained constantly, the resistance must be varied directly with the emf; on the other hand, if the emf is to be kept constant, the resistance must be varied inversely with the current.

Ohm's Law and Power

There is another series of equations related to Ohm's law because of the fact that the power in watts in any electrical circuit is equal to the current in amperes multiplied by the emf in volts. Thus:

$$P = EI$$

where

    P = power in watts

    I = current in amperes

    E = emf in volts

This formula may be transposed to find either I or E. Thus:

$$I = \frac{P}{E} \quad \text{and} \quad E = \frac{P}{I}$$

So if $P = 100$ watts and $I = 4$ amperes, we find $E = P/I$, or 100 W/4 A = 25 V. Since $R = E/I$ and $E = P/I$,

$$R = P/I^2, \quad I^2 = P/R \quad \text{and} \quad I = \sqrt{P/R}, \quad \text{and} \quad P = I^2R$$

Then, if $P = 100$ watts and $R = 4$ ohms, then $I^2 = 100$ W/4 Ω = 25 A^2, and the square root of 25 is 5, so $I = 5$ amperes. Further, since $P = EI$ and $I = E/R,$ we have:

$$E^2/R = P \quad R = E^2/P \quad E^2 = RP \quad \text{and} \quad E = \sqrt{RP}$$

If $R = 4$ ohms and $P = 100$ watts, then $E^2 = PR = 100$ W $\times$ 4 Ω = 400 V^2. The square root of 400 V^2 is 20 V. Thus, $E = 20$ volts.

The Ohm's Law Circle

One of the easiest ways to remember, learn, and use Ohm's Law is with the circle diagram shown in Figure 5-2. This Ohm's law circle can be used to obtain all three of these formulas easily. The method is this: Place your finger over the value that you want to find (E for voltage, I for current, or R for resistance), and then the other two will make up the formula. For example, if you place your finger over the E in the circle, the remainder of the circle will show $I \times R$. If you then multiply the current times the resistance, you will get the value for voltage in the circuit. If you wanted to find the value for current, you would put your finger over the I in the circle, and then the remainder of the circle will show $E \div R$. So, to find current we divide voltage by resistance. Lastly, if you place your finger over the R in the circle, the remaining part of the circle shows $E \div I$. These Ohm's law formulas apply to any electrical circuit, no matter how simple or how complex.

Ohm's Law

Voltage = Current × Resistance
Current = Voltage ÷ Resistance
Resistance = Voltage ÷ Current

Figure 5-2   The Ohm's Law Circle.

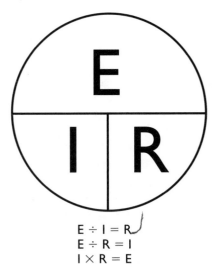

$E \div I = R$
$E \div R = I$
$I \times R = E$

Formulas
Ohm's Law

$$E = IR$$
$$I = E/R$$
$$R = E/I$$

Power in Watts

$$P = IE$$
$$I = P/E$$
$$E = P/I$$
$$R = P/I^2$$

$$I^2 = P/R$$
$$I = \sqrt{P/R}$$
$$P = I^2/R$$
$$P = E^2/R$$
$$R = E^2/P$$
$$E^2 = RP$$
$$E = \sqrt{RP}$$

Questions

1. State Ohm's law in simple form (in words).
2. State Ohm's law as an equation using the customary letter for each quantity. (Give the meaning of each letter used.)
3. Give three equations for Ohm's law for determining the values of current, emf, and resistance.
4. Give three equations for the power in an electrical circuit when it is desired to have watts, amperes, or voltage.
5. Give three equations for the power in an electrical circuit when the quantities involved are watts, current, and resistance.
6. Give the equations for the power in an electrical circuit when the quantities involved are voltage, watts, and resistance.
7. A load absorbs 650 watts. It is designed to operate at 100 volts. How many amperes will it draw?
8. A lamp has a resistance of 100 ohms and takes 1 ampere. How many watts does it absorb?
9. A coil is designed to take 5 amperes from a 110-volt circuit. What is its resistance and how many watts does it absorb?
10. A motor has an equivalent resistance of 6 ohms and absorbs 2.4 kW. How many amperes does it take? For what voltage is it designed?
11. A battery having an internal resistance of 4 ohms is connected in series with an external resistance of 6 ohms. A current of 5 amperes circulates. What is the voltage of the battery?
12. A current of 8 amperes must flow through a total resistance of 3 ohms. What is the voltage required?

Chapter 6

Capacitors

Capacitors are also termed condensers, especially in older texts and in the automobile industry. In this chapter, we will use the term capacitor, which is the most accepted terminology at the present time.

We found earlier that two oppositely electrified bodies attract one another. A capacitor is two electrified conducting surfaces separated by a dielectric.

A very practical method of describing capacitor action is shown in Figure 6-1. Two metal discs, A and B, are mounted on insulating supports. Disc B is connected by a conductor to a static machine or other high-voltage DC source, so that B becomes positively charged. Disc A is connected by a conductor to ground. Plate B will inductively charge the surface of plate A nearest to B negatively and an equal positive charge will go to ground through conductor G.

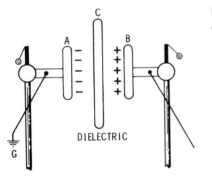

Figure 6-1 Principle of a capacitor.

The pith ball electroscope connected to A will show a feeble electrification, while the pith ball connected to B shows more electrification because B is connected to the static machine.

The negative charge on A reacts to draw the positive charge on B to B's inner surface. This allows more positive charge to flow from the static machine. The increased positive charge on B induces a stronger negative charge on A. This action continues until the dielectric C breaks down and a flash occurs between A and B.

Before the dielectric breaks down, if the static machine and ground are disconnected and the plates *A* and *B* pushed closer together, the pith balls will fall, which one might interpret to mean less charge is present. This, however, is not the case. What is happening is the capacitance of plates *A* and *B* increases as they approach, accompanied by a fall of potential.

When *A* and *B* are pulled apart, the capacitance decreases and the potential rises, causing the pith balls to fly outward due to the potential increase.

The arrangement just described is called a *capacitor*. The dielectric *C* may be air, paper, mica, glass, or any other dielectric. The plates *A* and *B* may be any metal, usually tin foil or aluminum foil. The capacitance of a capacitor depends upon three things:

1. The size and shape of the metal coatings
2. The thinness of the intervening dielectric
3. The dielectric separating the coatings

Please note that the material used for the metal coatings doesn't affect a capacitor's capacitance. The size and shape of the metal does affect the capacitance. The thinner the dielectric, the more capacitance the capacitor has. The material that composes the dielectric also affects its capacitance.

Using a vacuum or air as a base of 1 for dielectric constants, Table 6-1 shows a few of the dielectric constants.

Table 6-1 Dielectric Constants

Air	1	Mica	$5.7 - 7$
Glass, crown	$5 - 7$	Paper, dry	$2 - 2.5$
Glass, flint	$7 - 10$	Paraffin wax	$2 - 2.3$
India rubber	$2.1 - 2.3$	Water, pure	81

The Leyden jar was probably the first capacitor. It was named for the town of Leyden, where it was developed. A glass jar is covered partway up on the outside with tinfoil and lined inside with tinfoil to the same height, and this inner foil is connected to a brass ball outside the jar. See Figure 6-2. The two layers of foil are the two plates, and the glass jar is the dielectric. These foils were used with static generators, one electrode attached to the outside foil and the other electrode attached to the inside foil. The Leyden jar was charged and then the leads disconnected. The Leyden jar would often hold a charge for days, most of the time element of holding a

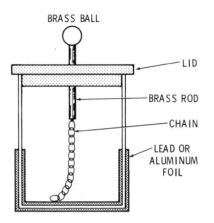

BRASS BALL

LID

BRASS ROD

CHAIN

LEAD OR
ALUMINUM
FOIL

Figure 6-2 Leyden jar capacitor.

charge depending upon the air's humidity. When the inside and out-side foil were electrically connected, an electric discharge would occur.

More than one Leyden jar could be connected together. In Figure 6-3, where the Leyden jars are connected in *series* (inner foil connected to outer foil of preceding jar) and the two discharge balls approach each other, the discharge wouldn't be heavy, but would discharge quite a distance, as the emf of each Leyden jar would add to each of those in series; thus, an increased potential would result.

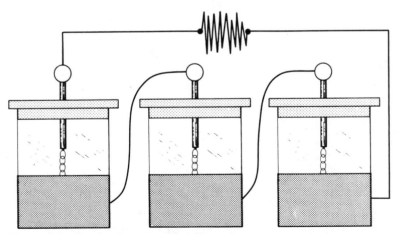

Figure 6-3 Leyden jars connected in series.

When Leyden jars are *paralleled* as in Figure 6-4 (inner foils connected together), the discharge emf would be the same as for each individual Leyden jar and the discharge gap would be only one-third of that in Figure 6-3, but the spark would be thicker due to the current of the Leyden jars adding up (the current would be three times as heavy as in Figure 6-3).

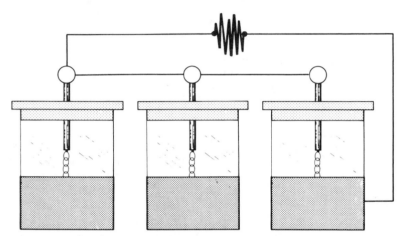

Figure 6-4 Leyden jars connected in parallel.

Leyden jars are really used only for experiments nowadays, but they very effectively illustrate the theory of capacitors.

The most commonly used form of capacitor is composed of layers of foil separated by wax paper as the dielectric, or some other similar dielectric. See Figure 6-5. The foil is cut into long narrow strips, separated by waxed paper, and then rolled up into a cylinder as shown in Figure 6-6. This of course is not the only form that capacitors appear in.

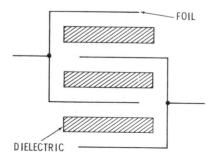

Figure 6-5 A commonly used capacitor in finished form (schematic).

Figure 6-6 A commonly used
form of capacitor.

A more recent type of capacitor is the electrolytic capacitor. The plates of this type of capacitor are polarized and marked anode (+) and cathode (−). The anode is the aluminum foil with large surface area, while the cathode is usually an aluminum container, but not necessarily so. There are both wet and dry types of electrolytic capacitors. The most common type is the wet type, with a solution of borax and boric acid in water. In another type, the electrolyte is ammonium citrate in water. The dry type uses a paste electrolyte of boric acid, glycerine, and ammonia. Paper or gauze packed between the anode and cathode is impregnated with the solution.

A DC source of emf is connected to the electrolytic capacitor, with the positive (+) to the anode and the negative (−) to the cathode. A slight current is established through the electrolyte, which produces a thin film on the surface of the anode, polarizing the capacitor. This film also serves as the dielectric. This is termed *forming* the capacitor.

If the polarity is reversed, the film will break down and the capacitor will be destroyed, often blowing up.

There is also an electrolytic capacitor that is not polarized; it is used in starting single-phase motors. These AC electrolytic capacitors can't be subjected to AC for long periods or too often in quick succession.

A capacitor doesn't pass electrons. (Lest one contest this statement, since there is no perfect insulator, we probably should state that *basically* a capacitor passes no electrons.)

As stated previously, the two plates of a capacitor become charged, one negative and one positive. A good way to demonstrate this is with a capacitor, a two-way switch, a DC source of emf, and a galvanometer. (We have not discussed galvanometers as yet, as these will be covered in the chapter on meters, but a galvanometer is a delicate meter for registering current.)

When the switch A in Figure 6-7 is closed, the galvanometer will show current as waveform B of Figure 6-8. The peak is the maximum flow point and then the flow tapers off. Thus the voltage doesn't all appear across the capacitor at first but must build up as the current drops. Then, when switch B in Figure 6-7 is closed, the galvanometer will show current as waveform C of Figure 6-8. The voltage across the capacitor drops to zero as the current goes to zero and the negative charges move to the other plate of the capacitor. Thus, a

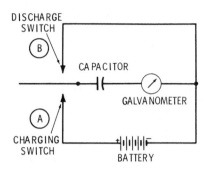

Figure 6-7   Charging and discharging a capacitor.

capacitor opposes changes in voltage. (The circuit property that opposes changes in current is called *inductance* and will be discussed later.)

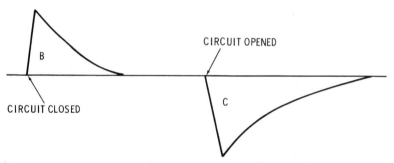

Figure 6-8   Current impulses of the circuit in Figure 6-7.

Figure 6-9 illustrates a capacitor in series with a light bulb and a DC source of emf from a battery. When switch A is closed, the lamp will light up for an instant and then go out. This is due to the current to charge the plates of the capacitor. When the capacitor is charged, no more current will flow.

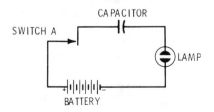

Figure 6-9   A lamp connected in series with a capacitor and a DC source.

Figure 6-10 shows a capacitor in an AC circuit, with a switch and lamp. When the switch is closed, the lamp will light and continue to stay lit. Actually, the current doesn't flow through the capacitor, but the plates are alternately charged positive and negative, giving the effect of current flowing through the capacitor. This happens so rapidly that the filament of the lamp doesn't have time to cool down, and thus it appears as if current flows steadily. During one cycle, the current flows first in one direction and then flows in the other direction. Thus, with 60 cycles, there are 120 alternations per second.

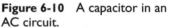

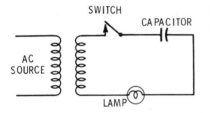

SWITCH
CAPACITOR
AC SOURCE
LAMP

Figure 6-10 A capacitor in an AC circuit.

Capacitance

The unit of capacitance is explained in Chapter 2 and is designated by F. A capacitor thus has a capacitance of one farad when one coulomb delivered to it will raise its potential one volt. The farad is a very large unit of measurement, so we usually talk in terms of microfarads (μF) or one one-millionth of a farad (10^{-6} farad).

Let Q be the charge of a capacitor in coulombs, E the applied potential difference in volts, and C the capacitance of a capacitor in farads. Then

$$\text{Capacitance} = \frac{\text{Charge}}{\text{Potential Difference}} \quad \text{or} \quad C = \frac{Q}{E}$$

As previously stated, the capacitance of a capacitor is affected by the material and dimensions of the dielectric. The capacitance is directly proportional to the effective area of the dielectric and inversely proportional to the thickness of the dielectric. Capacitance is also proportional to the dielectric constant (examples were given in Table 6-1) of the material between the plates.

Let A be the total area of dielectric between the plates, in square inches, s the thickness of dielectric in inches, κ the dielectric constant, and C the capacitance of the capacitor in microfarads. Then

$$C = \frac{2.24 \times k \times A}{s \times 10^7}$$

is the capacitance.

Now, using the dielectric constant (from Table 6-1) and measurements of the dielectric, we may find the capacitance of a capacitor. We have just stated that the capacitance is directly proportional to the effective area of the dielectric. It stands to reason that the total area of the dielectric will be larger than the plate area, for insulating purposes, so effective area of the dielectric is also the plate area.

Problem
A capacitor is composed of 31 tinfoil sheets measuring 6.5 inches by 6 inches, separated by mica 0.004 inch thick (4 mils) with a dielectric constant of 6.5. What is the capacitance of this capacitor?

There are 30 sheets of mica between the plates. The plates are 6.5 inches by 6 inches, so there are 39.0 square inches of effective dielectric per sheet and 30 sheets. Thus

$$C = \frac{2.24 \times k \times A}{s \times 10^7} = \frac{2.24 \times 6.5 \times 39 \times 30}{0.004 \times 10,000,000} = 0.43\,\mu\text{F}$$

Capacitance in Series and Parallel
When we parallel capacitors in a circuit, such as in Figure 6-11, we may have to find their combined capacitance.

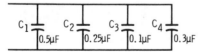

C_1 0.5μF C_2 0.25μF C_3 0.1μF C_4 0.3μF

Figure 6-11   Capacitors in parallel.

To find the combined capacitance of capacitors in parallel, add the capacitance of all of the individual capacitors in the circuit. Thus, in Figure 6-11

$C_1 = 0.50\ \mu F$

$C_2 = 0.25\ \mu F$

$C_3 = 0.10\ \mu F$

$\dfrac{C_4 = 0.30\ \mu F}{C\ = 1.15\ \mu F}$

The total capacitance will always be larger than that of any one capacitor in the circuit.

When we put capacitors in series in a circuit, as in Figure 6-12, we may have to find the total capacitance of the circuit, and this is more complicated. When capacitors are in series in a circuit, the total capacitance will be less than that of any individual capacitor in the circuit.

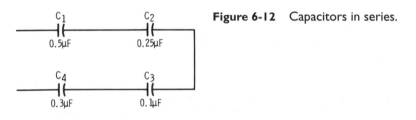

Figure 6-12   Capacitors in series.

We find the total capacitance by taking the reciprocal of the sum of the reciprocals of the individual capacitors. You may ask, "What is a reciprocal?" Here are some examples:

The reciprocal of 2 is ½.

The reciprocal of ½ is 2.

So what we actually do is invert the capacitance of the individual capacitors, as in Figure 6-12, into reciprocals or fractions and add them:

$C_1 = 0.50\ \mu F$:   Reciprocal $= 1/0.5\ \mu F$

$C_2 = 0.25\ \mu F$:   Reciprocal $= 1/0.25\ \mu F$

$C_3 = 0.10\ \mu F$:   Reciprocal $= 1/0.1\ \mu F$

$C_4 = 0.30\ \mu F$:   Reciprocal $= 1/0.3\ \mu F$

Now,

$$\frac{1}{C_t} = \frac{1}{C_1} + \frac{1}{C_2} + \frac{1}{C_3} + \frac{1}{C_4}$$

So,

$$\frac{1}{0.5} + \frac{1}{0.25} + \frac{1}{0.1} + \frac{1}{0.3}$$
$$= \frac{3.0}{1.50} + \frac{6.0}{1.50} + \frac{15.0}{1.50} + \frac{5.0}{1.50} = \frac{29.0}{1.50}$$

This is the sum of the reciprocals. Now the reciprocal of the sum of the reciprocals will be $1.50/29.0 = 0.051$ μF, which is the answer and it is, as stated, less than the capacitance of the smallest capacitor in the circuit.

Capacitance in Other Than Regular Capacitors

We have discussed capacitors and capacitance as it pertains to regular capacitors. It would be amiss to in any way leave the impression that capacitance appears only in capacitors as such.

As we progress further, we will find capacitance entering into our discussions in many ways. For now, let us state that if we have conductors in a raceway (metallic), there will be capacitor effect between the conductors and between the conductors and grounded raceway. See Figure 6-13. In a later chapter on insulation testing, you will be cautioned after running insulation tests to short-out the conductors and the raceway. This is because of the charge accumulated in the capacitive effect, from which a severe shock may result. Coaxial high-voltage cables create a capacitor charge between the shield of the cable and the conductors, which may result in a very high voltage discharge.

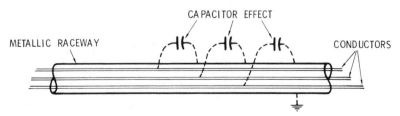
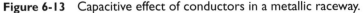

Figure 6-13 Capacitive effect of conductors in a metallic raceway.

Power lines on poles have a capacitive effect between conductors and conductors and ground. We have not as yet touched on power factor and inductive loads, but it may not be wrong to mention a case where we served a town 40 miles away from the power plant by a 44,000-volt transmission line. The load in the town was, of course, slightly inductive (opposing changes of current) in nature and usage in that era was light compared with that of this day and age. The line and load at the power plant showed a predominance of capacitance (opposing change of voltage). This was due to the capacitance of the line being greater than the inductance of the load. Of course, this changed as the load in the town built up.

Finally, you should observe that whenever you have occasion to contact both terminals of a capacitor of any size, always short-out the two terminals before touching them, to discharge the capacitor, lest you receive a severe shock.

Formulas

$$C = \frac{Q}{E} \quad \text{or} \quad \text{Capacitance} = \frac{\text{Charge}}{\text{Potential Difference}}$$

$$C = \frac{2.24 \times k \times A}{s \times 10^7}$$

Capacitors in parallel:

$$C_1 + C_2 + C_3 + C_4 + \cdots = \text{Total } C \text{ or } C_t$$

Capacitors in series:

$$C_t = \frac{1}{C_1} + \frac{1}{C_2} + \frac{1}{C_3} + \cdots + \frac{1}{C_n}$$

Questions

1. Define (a) potential, (b) capacitance.
2. What is the relation of electrical charge to electrical capacitance and potential? Express this by mathematical formula and in your own words.
3. What are the necessary parts of a simple capacitor?
4. Upon what three things does the capacitance of a capacitor depend? Explain how a change in each affects the capacitance of a capacitor.

5. A body possesses a fixed quantity of charge. If the capacitance is increased, what will be the effect on the potential?

6. Where do the charges reside in a capacitor? Give proof.

7. What is the purpose of metal plates in a capacitor?

8. Describe the construction of a Leyden jar.

9. Mention a number of the better dielectrics.

10. Explain the construction of any standard form of a capacitor. What materials are used and what are their advantages?

11. A capacitor is made up of 54 sheets of tinfoil, each 10 inches long and 6 inches wide, which are separated by sheets of waxed paper of dielectric constant of 2.25 and have a thickness of 0.008 inch. Alternate sheets of tinfoil are connected to one terminal of the capacitor and the remaining sheets of tinfoil are connected to the other terminal. Find the capacitance of the capacitor. (Ans: 0.22 μF.)

Chapter 7

Resistance

In the latter part of Chapter 1 a short discussion on insulators and conductors was given. It was stated that there is no perfect insulator. Also, a few metals in the order of their conductivity appeared, if you wish to refer back to this part.

The word resistance comes from resist, meaning to oppose. We may compare it to the friction of liquids flowing in pipes (see Figure 7-1). We have two water pipes, one ½ inch in diameter and one 1 inch in diameter, both connected to the same source of water and both receiving the same water pressure. The flow will be far less out of the ½-inch pipe than out of the 1-inch pipe due to more friction (restriction) in the smaller pipe. The basic principle is the same for conductors and resistance: Smaller conductors have more resistance and less current-carrying capacity than larger conductors of the same material.

Figure 7-1 Analogy of friction and resistance.

The electrical current through a conductor will have a loss of power and a drop of voltage. The current value squared times the ohmic resistance (I^2R loss) is the power loss, and the current (I) times the resistance (R) is IR, or the drop in potential, which are both the results of current flowing through resistance. (Refer to Chapter 2 for definitions.) The unit of resistance is referred to as an ohm (Ω). Conductivity (G) is in mhos, which is the reciprocal of the resistance unit ($1/R = G$).

Thus, if a circuit has a resistance of 15 ohms, its conductivity is the reciprocal of 15 ohms, or 1/15 mho.

When current flows through resistance, heat is produced in the form of I^2R or power loss in watts. In many cases this is a loss that we attempt to keep to a minimum, while in other cases we make use of this loss.

In an electric heater, high-resistance wires are used on purpose to get a high I^2R loss or heat. In the incandescent lamp, high-resistance filaments are used, which heat to a white heat-producing light.

In circuits for carrying electrical currents to the point of use, the resistance should be kept low, as our purpose is to use the electrical current at the far end of the circuit and any losses in the conductors in this circuit, whether IR drops or I^2R losses, are purely losses. One could think of, say, the cost of carrying the current from one point to another.

Again, in electronic circuits, resistances are used to drop voltages purposely or to limit currents. From this, one may readily see that resistance may be useful or detrimental.

In Chapter 2 the definition of an ohm is given. Also, prefixes such as micro, milli, mega, etc., are also given. You may refer back to these, as terms such as megohms, milliamperes, etc., will often be referred to as we progress.

Many factors affect the resistance of a conductor. Here are the primary ones:

1. The resistance of a conductor is directly proportional to its length; thus, using R_1 and L_1 for one conductor and R_2 and L_2 for the other conductor, you will get a proportion such as $R_1 : R_2 = L_1 : L_2$. For example, suppose $R_1 = 10$ ohms, $R_2 = 20$ ohms, $L_1 = 200$ ft, and $L_2 = x$. Then

$$10 : 20 = 200 : x$$

From arithmetic, the product of the means equals the product of the extremes. So,

$$20 \times 200 = 10x$$
$$4000 = 10x$$
$$x = 400 \text{ ft}$$

2. The resistance of a conductor is inversely proportional to its cross-sectional area. Using R_1 and A_1 for one conductor and R_2 and A_2 for the other conductor, we have $R_1 : R_2 = A_2 : A_1$. Shortly you will find that the cross-sectional area of a round conductor is proportional to the square of the diameter, so we will substitute d^2 (diameter squared) for A (cross-sectional area). Thus,

$$R_1 : R_2 = d_2^2 : d_1^2$$

For example, if $R_1 = 4$ ohms, $R_2 = x$, $d_2 = 0.20$ inch, and $d_1 = 0.05$ inch, then

$$4 : x = 0.20^2 : 0.05^2$$
$$4 : x = 0.04 : 0.0025$$
$$0.04x = 0.01$$
$$x = 0.25 \text{ ohm}$$

3. The resistance of a conductor of a given length and cross-sectional area depends upon the material of which it is composed. Thus, iron has approximately six times the resistance of copper; so if both the copper conductor and the iron conductor are the same cross-sectional area and length, the iron will have six times the resistance of the copper.

4. Temperature affects resistance. This has several aspects. All pure metals increase in resistance with an increase in temperature, but this increase is not linear. With copper the resistance increases approximately ⅕ of 1 percent for each degree rise in Fahrenheit above room temperature.

There are alloys that don't change resistance much with changes in temperature. These are said to have a *zero temperature coefficient.*

There are materials available, such as carbon, liquids, and insulation materials, that have a *negative* temperature coefficient.

In dealing with resistances, the term ohm/CM-foot will be of interest. This is read "ohms per circular-mil foot."

At this point a circular mil will be defined. A *mil* is defined as 1/1000 of an inch. A *circular mil,* therefore, can be defined as being the area of a circular conductor that has a diameter of 1 mil or 1/1000 inch. It is quite important to note this definition because in discussions concerning conductors, the term circular mils will be used quite frequently.

If the diameter of a round conductor is 0.003 inch, or 3 mils, or 3/1000 inch, then the cross-sectional area is $3 \times 3 = 9$ circular mils. Again, if a round conductor is 25 mils or 25/1000 inch in diameter, then its area is $25 \times 25 = 625$ circular mils. Thus, the cross-sectional area of a round conductor in circular mils is equal to the square of the diameter in mils.

See Table 7-1, covering the American Wire Gauge (AWG), which is sometimes termed B&S Gauge.

A round conductor 3 mils in diameter will be $3 \times 3 = 9$ circular mils in cross-section. Figure 7-2 illustrates the difference between square mils and circular mils. Here there are nine round conductors

Table 7-1   American Wire Gauge for Solid
Copper Conductors

| Size AWG | Wire Diameter (In) | Cross-Sectional Area | | DC Resistance: Soft Copper Max. Res. per 1000 Ft at 20°C (Ohms per 1000 Ft) |
		(Circular Mils)	(Square-Inch)	
4/0	0.4600	211,600	0.1662	0.04993
3/0	0.4096	167,800	0.1318	0.06296
2/0	0.3648	133,100	0.0145	0.07939
1/0	0.3249	105,400	0.08289	0.1001
1	0.2893	83,690	0.06573	0.1262
2	0.2576	66,379	0.05213	0.1592
3	0.2294	52,634	0.04134	0.2007
4	0.2043	41,472	0.03278	0.2531
5	0.1819	33,100	0.02600	0.3192
6	0.1620	26,250	0.02062	0.4025
7	0.1443	20,820	0.01635	0.5075
8	0.1285	16,510	0.01297	0.6400
9	0.1144	13,090	0.01028	0.8070
10	0.1019	10,380	0.008155	1.018
11	0.09074	8234	0.006467	1.283
12	0.08081	6530	0.005129	1.618
13	0.07196	5178	0.004067	2.040
14	0.06408	4107	0.005225	2.573
15	0.05707	3257	0.002558	3.244
16	0.05082	2583	0.002028	4.091
17	0.04526	2048	0.001609	5.158
18	0.04030	1624	0.001276	6.505
19	0.03589	1288	0.001012	8.202
20	0.03196	1022	0.0008023	10.34
21	0.02846	810.1	0.0006363	13.04
22	0.02535	642.5	0.0005046	16.45
23	0.02257	509.5	0.0004001	20.74
24	0.02010	404.0	0.0003173	26.15
25	0.01790	320.4	0.0002517	32.97
26	0.01594	254.1	0.0001996	41.58
27	0.01420	201.5	0.0001583	52.43
28	0.01264	159.8	0.0001255	66.11

Table 7-1 *(continued)*

Size AWG	Wire Diameter (In)	Cross-Sectional Area (Circular Mils)	(Square-Inch)	DC Resistance: Soft Copper Max. Res. per 1000 Ft at 20°C (Ohms per 1000 Ft)
29	0.01126	126.7	0.00009954	83.37
30	0.01003	100.5	0.00007894	105.1
31	0.008928	79.70	0.00006260	132.6
32	0.007950	63.21	0.00004964	167.2
33	0.007080	50.13	0.00003937	210.8
34	0.006305	39.75	0.00003122	265.8
35	0.005615	31.52	0.00002476	335.2
36	0.005000	25.00	0.00001963	422.6
37	0.004453	19.83	0.00001557	532.9
38	0.003965	15.72	0.00001235	672.0
39	0.003531	12.47	0.000009793	847.4
40	0.003145	9.888	0.000007766	1069

Figure 7-2 Comparative illustration of 9 circular mils and 9 square mils.

├──────── 3 MILS ────────┤

of 1 mil diameter all in a square of 3 mils per side, or 9 square mils, with the circle inscribed within the square having a diameter of 3 mils or an area of 9 circular mils. Thus, obviously, 9 square mils is larger than 9 circular mils.

The nine small 1-mil circles touch the square, but there are voids created. The area of a circle is πr^2, or $\pi d^2/4$ (diameter squared).

Note that $\pi d^2/4 = 0.7854d^2$. Thus, the area of the nine small circles is $0.7854 \times (1 \text{ mil})^2 \times 9 = 7.0686$ square mils. From this, the area of the large circle is equal to the total area of the nine small circles, and you also see that the area of a circle 3 mils in diameter is 0.7854 of the square that is 3 mils per side.

In order to convert a unit of circular-mil area into its equivalent area in square mils, the circular-mil area must be multiplied by $\pi/4$ or 0.7854, which is the same as dividing by 1.273. Conversely, to convert a unit area of square mils into its equivalent in circular mils, the square mils should be divided by 0.7854, which is the same as multiplying by 1.273. The above relations may be written as follows:

$$\text{Square Mils} = \text{Circular Mils} \times 0.7854 = \frac{\text{Circular Mils}}{1.273}$$

$$\text{Circular Mils} = \frac{\text{Square Mils}}{0.7854} = \text{Square Mils} \times 1.273$$

See Figure 7-3.

From the above, a rectangular area such as a bus bar may have its area computed in square mils and then an easy conversion may be made to circular mils to arrive at current-carrying capacities (ampacities).

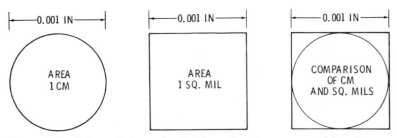

Figure 7-3 Enlarged view of a circular mil and a square mil for direct comparison.

A *circular mil-foot* is the cylindrical conductor that is 1 foot in length and 1 mil in diameter. The resistance of such a unit of copper has been found experimentally to be 10.37 ohms at 20°C; this is normally considered to be 10.4 ohms. See Figure 7-4. A circular mil-foot of copper at 20°C offers 10.4 ohms' resistance; at 30°C, it is 11.2 ohms; at 40°C, it is 11.6 ohms; at 50°C, it is 11.8 ohms; at

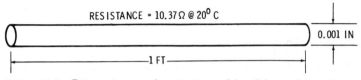

Figure 7-4 Dimensions and resistance of 1 mil-foot of round copper conductor.

60°C, it is 12.3 ohms; and at 70°C, it is 12.7 ohms. In voltage-drop calculations (*IR* loss), the quantity 12 ohms is very often used as the constant; so, unless otherwise specified, this is the constant that we will be using. It takes into consideration higher temperatures and thus gives a safety factor, such as for resistance of connections.

Table 7-2 gives ohms per circular mil-foot of a number of conductors. From Table 7-2 it is found that at 20°C the resistance of 1 circular-mil foot of annealed copper is 10.4; of aluminum, it is 17.0; of iron, 60.2; and of Nichrome, 660. From this it is found that aluminum has 1.63 times the resistance of copper, all other considerations being equal. Iron has six times the resistance of copper, and Nichrome has 63.5 times the resistance of copper.

Table 7-2 Specific Resistances (Resistivities) of Various Conductors at 20°C (68°F)

Conductor	Ohms per Circular-Mil Foot	Ohm-Cm $\times$ 10^{-6}
Aluminum	17.0	2.62
Carbon (graphite approx.)	4210.0	1400
Constantan (Cu 60%; Ni 40%)	295.0	44.2
Copper (annealed)	10.4	1.7241
Iron (99.98% pure)	60.2	9.71
Lead	132.0	21.9
Manganin (Cu 84%; Ni 4%; Mn 12%)	264.0	44
Mercury	576.0	21.3
Nichrome (Ni 60%; Cr 12%; Fe 26%; Mn 2%)	660.0	100
Platinum	59.5	10.5
Silver	9.9	1.62
Tungsten	33.1	5.48
Zinc	36.7	6

This gives a better understanding of why Tables 310.16 through 310.31 in the *National Electrical Code* require larger sizes of aluminum conductors than copper conductors of the same ampacity (current-carrying capacity).

Indeed, the resistance R of a cylindrical conductor is proportional to its length L and inversely proportional to its cross-sectional area A. This may be written as

$$R = \rho \frac{L}{A}$$

where ρ (Greek letter "rho") is the coefficient of proportionality and is called the *resistivity* of the material. Solving the above equation for ρ, we get $\rho = RA/L$, which may be written as follows:

$$\rho = \frac{R}{L/A}$$

Thus, resistivity is expressed in ohms per unit length per unit area. Here is where the term "ohms per circular-mil foot" is used for resistivity ρ, when R is given in ohms, L in feet, and A in circular mils.

When the length L is the one-way distance for both the lead and return conductors, the following formula must be used:

$$R = \frac{2\rho L}{A}$$

This is because the total length of conductor must be employed. This formula is for DC and single-phase AC circuits.

Example

What is the resistance of a 500-ft circuit of No. 4 copper conductor? Take $\rho = 12$ for copper and $\rho = 18$ for aluminum.

In this case, A = 41,472 CM (from Table 7-1) and $\rho = 12$. So

$$R = \frac{2 \times 12 \times 500}{41,742} = 0.29 \text{ ohm}$$

is the resistance of the circuit.

The preceding formula for resistance may be manipulated to give the cross-sectional area:

$$A = \frac{2\rho L}{R}$$

where A is usually expressed in circular mils.

Example

Suppose it is desired to have a copper conductor of 0.5 ohm resistance, the total length of which is 2000 feet (1000 feet one way). What must be the total cross-sectional area? What size conductor will be required?

From the preceding formula,

$$A = \frac{2\rho L}{R} = \frac{2 \times 12 \times 1000}{0.5} = 48,000 \text{ CM}$$

From Table 7-1, No. 4 wire has a cross-sectional area of 41,472 CM and No. 3 has a cross-sectional area of 52,634 CM. Since 41,472 is less than 48,000, No. 3 AWG must be selected.

The preceding formula may be transposed to solve for the length *L*:

$$L = \frac{RA}{2\rho}$$

where *L* is the conductor length in feet, one way.

Notice that Table 7-2 also gives resistivities in ohm-centimeters times 10^{-6}. This is simply for calculations in which *L* and *A* are expressed in centimeters. We won't be using calculations such as these, but they are used to some extent in other literature and this table may be of value for that purpose.

Skin Effect

AC calculations have not been covered, but for future reference, when AC flows through a conductor, an inductive effect occurs, which tends to force the current to the surface of the conductor. This produces an additional voltage loss and also affects the current-carrying capacity of the conductor. For open wire or wires in nonmetallic-sheathed cable or metallic raceways, the skin effect is negligible until size No. 2 AWG is reached. At this point the *NEC* gives multiplying factors for conversion from DC to AC resistance. Please note that there is a different factor for copper and aluminum conductors. See Table 7-3.

Conductivity

Conductivity is the ability of a material to conduct currents. It is the reciprocal of resistivity. Table 7-4 lists the conductivity of some metals, with silver taken as 100 percent conductive.

Table 7-3   (Table 9 of the *NEC*) Multiplying Factors for Converting DC Resistance to 60-Hz AC Resistance

	Multiplying Factor			
	For Nonmetallic Sheathed Cables in Air or Nonmetallic Conduit		For Metallic Sheathed Cables or all Cables in Metallic Raceways	
Size	Copper	Aluminum	Copper	Aluminum
Up to 3 AWG	1	1	1	1
2 AWG	1	1	1.01	1.00
1 AWG	1	1	1.01	1.00
0 AWG	1.001	1.000	1.02	1.00
00 AWG	1.001	1.001	1.03	1.00
000 AWG	1.002	1.001	1.04	1.01
0000 AWG	1.004	1.002	1.05	1.01
250 MCM	1.005	1.002	1.06	1.02
300 MCM	1.006	1.003	1.07	1.02
350 MCM	1.009	1.004	1.08	1.03
400 MCM	1.011	1.005	1.10	1.04
500 MCM	1.018	1.007	1.13	1.06
600 MCM	1.025	1.010	1.16	1.08
700 MCM	1.034	1.013	1.19	1.11
750 MCM	1.039	1.015	1.21	1.12
800 MCM	1.044	1.017	1.22	1.14
1000 MCM	1.067	1.026	1.30	1.19
1250 MCM	1.102	1.040	1.41	1.27
1500 MCM	1.142	1.058	1.53	1.36
1750 MCM	1.185	1.079	1.67	1.46
2000 MCM	1.233	1.100	1.82	1.56

Table 7-4 Conductivity of Metals

Silver	100%	Iron	16%
Copper	98%	Lead	15%
Gold	78%	Tin	9%
Aluminum	61%	Nickel	7%
Zinc	30%	Mercury	1%
Platinum	17%		

A statement was previously made that iron has six times the resistance of copper, which is arrived at by dividing 16 percent (iron) into 98 percent (copper). With aluminum, divide 61 percent into 98 percent (copper) and you find that aluminum has 1.6 times the resistance of copper.

There is another item—*superconductivity*—that would be good to remember. Certain metals show a remarkable drop in resistance at very low temperatures. This condition is known as superconductivity.

A Dutch physicist, H. Kamerlengh Omnes, as early as 1914 reported a 1-hour test in which current was produced in a lead ring at a very low temperature, and then the source of emf was removed. The current in the lead ring continued without appreciable reduction throughout the test. It was found that at $-266°C$, the resistance of lead is less than 10^{-12} of its value at 20°C. This is 1/1,000,000,000,000 of its resistance at 20°C.

This phenomenon has been observed in other metals and alloys. It is not our intent to put special importance on superconductivity, but findings are often experimental at the time of discovery, and then play a very important role later in the applications of use.

Voltage-Drop Calculation

In previous discussions it was found that IR drop (voltage) in circuits is sometimes detrimental and results in I^2R loss (power), resulting in heating that is not put to any real use.

Formulas will be given to calculate voltage drop. There are many tables available for calculating voltage drop, but if the basic theory is mastered, then these tables will have more meaning to you.

In Chapter 5, it was stated that $E = IR$. In Table 7-1, the circular-mil area of the various sizes of AWG conductors is listed. This table is not always at your fingertips and the circular-mil area of conductors is used in most of the following formulas. Therefore, it is suggested that the CM area of No. 10 be remembered: It is 10,380 CM. From this figure the entire table of CM area is at your fingertips, with a little effort, i.e., wire three sizes removed doubles or halves in area. Thus, No. 7 is 20,800 CM, or double that of No. 10, and No. 13 is 5180 CM, or half that of No. 10. Ten sizes removed is 10 times the area or $\frac{1}{10}$ of the area. Thus, No. 20 is 1020 CM, or $\frac{1}{10}$ that of No. 10, and No. 1/0 is 105,400 CM or 10 times that of No. 10. Granted that these figures are not exactly the same as in Table 7-1, but they are close enough for all practical purposes.

The *National Electrical Code* tells us (see Audel's *Guide to the NEC*) that 3 percent is the allowable voltage drop for feeds and

5 percent is the maximum voltage drop for the combination of feeder and branch circuits.

We may calculate voltage drops in conductors by taking the equation $R = 2\rho L/A$ and multiplying by I:

$$E_d = IR = \frac{2\rho LI}{A}$$

and

$$A = \frac{2\rho LI}{E_d}$$

Example

A certain motor draws 22 amperes at 230 volts and the feeder circuit is 150 feet in length. If No. 10 copper conductors are selected, what would the voltage drop be? Would No. 10 conductors be permissible for the 3 percent maximum voltage drop allowed by the *NEC*?

$$E_d = \frac{2 \times 12 \times 150 \times 22}{10,380} = \frac{79,200}{10,380} = 7.63 \text{ volts}$$

However, $230 \times 0.03 = 6.90$ volts, which is the voltage drop permitted by the *NEC* for feeders, so No. 10 conductors wouldn't be large enough. The next larger size would be the proper selection.

In the Appendix, you will find a table to determine volt loss in copper conductors in iron and nonmagnetic conduits; you will also find a similar table for aluminum conductors.

Measuring Conductors

In measuring the diameter of conductors, there are two very popular instruments used. One is the wire gauge, illustrated in Figure 7-5. The other is the micrometer (Figure 7-6), which measures one-thousandths of inches and fractions thereof.

Many conductors are composed of stranded conductors. The total of the circular mils of the strands equals the total circular mils of the conductor. Conductors are stranded mostly for flexibility and ease of pulling into raceways. This is especially true of the larger sizes of conductors, such as No. 8 and larger.

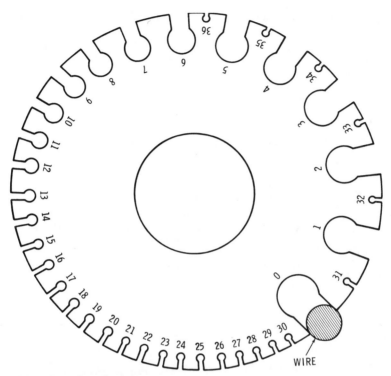

Figure 7-5 Standard wire gauge.

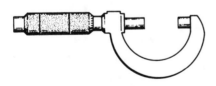

Figure 7-6 Micrometer.

Questions

1. What losses occur in conductors when electrical currents flow through them?
2. What is an ohm?
3. What is a mho?
4. How many ohms in a megohm?

5. The resistance of a conductor is directly proportional to its
_____.

6. The resistance of a conductor is inversely proportional to its
_____.

7. Does temperature affect the resistance of copper conductors?

8. What is a circular mil?

9. Give a formula for changing circular mils to square mils.

10. Give a formula for changing square mils to circular mils.

11. What is designated by rho?

12. Give the symbol for rho.

13. What is skin effect?

14. Give the conductivity of the following metals, as compared to silver: copper, gold, Nichrome, aluminum, and iron.

15. Explain superconductivity.

16. Give the voltage-drop formula for DC circuits.

Chapter 8

Resistance in Series and Parallel

A series circuit is one in which all the electrical devices and components are connected end to end so that the same current flows throughout the circuit. Such an electrical circuit is illustrated in Figure 8-1. In Figure 8-1, there are five resistors connected in series with a source of electrical energy, which in this case will be shown as the battery B.

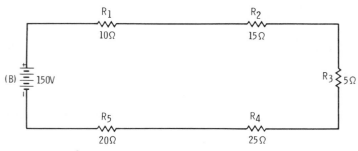

Figure 8-1   Series resistive circuit.

Resistances in Series

There are rules governing resistances in series, which are as follows:

1. The total resistance in a series circuit is equal to the arithmetic sum of the resistances of the individual resistors in the series circuit; e.g., in Figure 8-1, we have five resistors in series, namely, R_1, R_2, R_3, R_4, and R_5. The resistance of the individual resistors are $R_1 = 10$ ohms, $R_2 = 15$ ohms, $R_3 = 5$ ohms, $R_4 = 25$ ohms, and $R_5 = 20$ ohms. Thus, to arrive at the total resistance of this series circuit, we add all five individual resistances: $10 + 15 + 5 + 25 + 20 = 75$ ohms, that is, $R_t = 75$ ohms.

2. The voltage drop in a series circuit equals the sum of the voltage drops of the resistors in series, and this in turn equals the voltage of the supply source. Also, the sum of the voltage drops across the individual resistors will be the same as the voltage drop across all resistors in the series circuit.

3. The current in a series circuit is the same throughout the entire circuit.

Now for rules 2 and 3, make use of Ohm's law: $E = IR$. It was found in (1) that $R_t = 75$ ohms and the source voltage is given as 150 volts, so $I = E/R$, or $I = 150/75 = 2$ amperes, and this 2 amperes flows through all five resistors. Further, the voltage drop in each resistor is as follows:

$$E_1 = I \times R_1 = 2 \times 10 = 20 \text{ volts}$$
$$E_2 = I \times R_2 = 2 \times 15 = 30 \text{ volts}$$
$$E_3 = I \times R_3 = 2 \times 5 \ = 10 \text{ volts}$$
$$E_4 = I \times R_4 = 2 \times 25 = 50 \text{ volts}$$
$$E_5 = I \times R_5 = 2 \times 20 = \underline{40 \text{ volts}}$$
$$\text{total} = \overline{150 \text{ volts}}$$

This proves that the total voltage drop is equal to the sum of the voltage drops across the resistors and this total must equal the voltage of the electrical source.

Resistances in Parallel

A *parallel* circuit is also known as a *multiple circuit* or a *shunt circuit*. Figure 8-2 illustrates such a circuit.

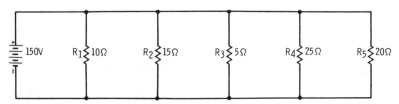

Figure 8-2 Parallel resistive circuit.

Again, as for series circuits, we have rules governing parallel circuits:

1. The total resistance of the combined resistances in a parallel circuit is always less than the resistance of the lowest-value resistor in the parallel circuit. This will be calculated mathematically as proof.

2. The combined total resistance of a number of unequal resistances in parallel is equal to the reciprocal of the sum of the reciprocals of the individual resistances.

We discussed the reciprocal of the sum of the reciprocals in Chapter 6 in studying capacitors. In Chapter 7 it was shown that conductance (mhos) was the reciprocal of resistance, so the total resistance of parallel resistors will equal the reciprocal of the sum of the conductances of the resistors in the parallel circuit.

Referring to Figure 8-2: R_1 = 10 ohms; R_2 = 15 ohms; R_3 = 5 ohms; R_4 = 25 ohms; R_5 = 20 ohms. Now, to find the total resistance of the parallel circuit, we take the reciprocal of each resistance (the conductance G) and add these; thus,

$$\frac{1}{10} + \frac{1}{15} + \frac{1}{5} + \frac{1}{25} + \frac{1}{20}$$

$$= \frac{30}{300} + \frac{20}{300} + \frac{60}{300} + \frac{12}{300} + \frac{15}{300} = \frac{137}{300}$$

Now the reciprocal of this answer is 300/137 = 2.189 ohms, which is the total resistance of this circuit. Note that the total resistance of a parallel circuit is less than the resistance of the smallest resistor in the circuit. (In this case the smallest resistor is 5 ohms.)

3. The voltage drop across each resistor in a parallel circuit is the same as the supply voltage. The supply voltage in this case is 150 volts, so each resistor receives 150 volts.

4. The total current (I_t) supplied by the source is equal to the supply voltage divided by the total resistance (R_t). In this circuit we found R_t = 2.189 ohms. Now let us check this, as the sums of the current in the resistors should equal I_t:

$$I_1 = \frac{E}{R_1} = \frac{150}{10} = 15 \text{ amperes}$$

$$I_2 = \frac{E}{R_2} = \frac{150}{15} = 10 \text{ amperes}$$

$$I_3 = \frac{E}{R_3} = \frac{150}{5} = 30 \text{ amperes}$$

$$I_4 = \frac{E}{R_4} = \frac{150}{25} = 6 \text{ amperes}$$

$$I_5 = \frac{E}{R_5} = \frac{150}{20} = 7.5 \text{ amperes}$$

$$\overline{\phantom{I_5 = \frac{E}{R_5} = } 68.5 \text{ amperes total}}$$

The total resistance divided into the source voltage should give us the same current or amperage:

$$I_t = \frac{E}{R_t} = \frac{150}{2.189} = 68.5 \text{ amperes}$$

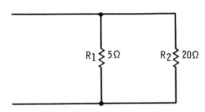

Figure 8-3 Calculation of two unequal resistances in parallel.

5. The combined resistance of two unequal resistances in parallel is equal to their product divided by their sum. Using Figure 8-3,

$$\frac{R_1 \times R_2}{R_1 + R_2} = \frac{5 \times 20}{5 + 20} = \frac{100}{25} = 4 \text{ ohms}$$

Checking this by the method in Rule 2: $1/5 + 1/20 = 20/100 + 5/100 = 25/100$, and the reciprocal of this fraction is $100/25 = 4$ ohms.

6. The combined resistance of any number of equal resistances in parallel may be found by dividing the resistance in one branch by the number of branches. See Figure 8-4.

For the circuit in Figure 8-4, $10/5 = 2$ ohms' total resistance. Again, checking this by Method 2: $1/10 + 1/10 + 1/10 + 1/10 + 1/10 = 5/10$, and the reciprocal of $5/10$ is $10/5 = 2$ ohms.

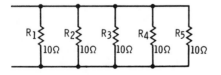

Figure 8-4 Calculation of equal resistances in parallel.

Series-Parallel Circuits

The next topic of discussion is the combination of series and parallel resistances. Refer to Figure 8-5.

The first step is to analyze the circuit and see what parts may be combined to simplify the circuit. In Figure 8-5 the two branches are indicated as A and B, respectively.

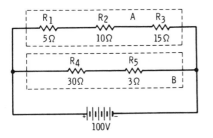

Figure 8-5 Series-parallel connection.

Take A first. Here we have R_1, R_2, and R_3 in series, so we add the resistances: $R_A = R_1 + R_2 + R_3 = 5 + 10 + 15 = 30$ ohms total in branch A. Also, $I = E/R = 100/30 = 3.33$ amperes in branch A. Now take branch B and find its total resistance: $R_4 + R_5 + R_B$ or $30 + 3 = 33$ ohms total in branch B. Now, $I = E/R = 100/33 = 3.03$ amperes in branch B. This gives us 3.33 amperes in A plus 3.03 amperes in B, or 6.36 amperes total.

We apply 100 volts and $R_t = E/I_t = 100/6.36 = 15.71$ ohms is the total resistance of the entire circuit in Figure 8-5.

Now, to check ourselves, $R_A = 30$ ohms and $R_B = 33$ ohms, so

$$\frac{R_A \times R_B}{R_A + R_B} = \frac{30 \times 33}{30 + 33} = \frac{990}{63} = 15.71 \text{ ohms}$$

In Figure 8-6 different portions of the circuit have been enclosed within dashed lines and lettered for ease in following calculations.

First, take A, which is three resistors in parallel, and find the total resistance of A: $1/15 + 1/10 + 1/3 = 2/30 + 3/30 + 10/30 = 15/30$, so $30/15 = 2$ ohms R_A total.

Next, R_A is in series with R_C, so $R_A + R_C = 2 + 5 = 7$ ohms, which is for branches A and C combined.

Next take branch B: $25 + 15 + 40$ ohms, which is the total for B. Now we have the A-C combination in parallel with B, so since A-C is 7 ohms and B is 40 ohms,

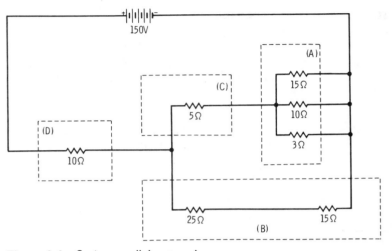

Figure 8-6   Series-parallel network.

$$\frac{R_{A-C} \times R_B}{R_{A-C} + R_B} = \frac{7 \times 40}{7 + 40} = \frac{280}{47} = 5.96 \text{ ohms total}$$

for the combination of A, B, and C.

Now D is in series with this combination, so $10 + 5.96 = 15.96$ ohms as the total resistance of the entire circuit in Figure 8-6.

The total current is $I = E/R$, so $I = 150/15.96 = 9.40$ amperes from the source. This indicates that the resistance in D carries 9.4 amperes and this divides between branches A, C, and B. The voltage drop across D is $E = IR = 9.4 \times 10 = 94$ volts. Then 150 volts – 94 volts = 56 volts as the voltage drop across the A, C, and B combination, or 56 volts drop across A-C, and the same 56 volts drop across B.

To check these figures for current values,

$$\text{In } A\text{–}C, I = E/R_{A-C} = 56/7 = 8 \quad \text{amperes}$$

$$\text{In } B, I = E/R_B = 56/40 = \underline{1.4 \text{ amperes}}$$
$$\overline{9.4 \text{ amperes}}$$

This is the same value of current that was found for D—the answers check.

Questions

1. Define a series circuit. Sketch one.

2. Define a parallel circuit. Sketch one.

3. Give a rule for combined resistance in a series circuit. What is the combined resistance of 50, 100, 10, 40, and 60 ohms in series?

4. Give the rule for combining resistance of two unequal resistances in parallel. What is the combined resistance of 30 and 50 ohms in parallel?

5. Give the rule for the combined resistance of any number of equal resistances in parallel. What is the combined resistance of 15 lamps each of 105 ohms' resistance in parallel?

6. Give a rule for the combined resistance of any number of unequal resistances in parallel. What is the combined resistance of 10-, 12-, 24-, 6-, and 8-ohm resistors in parallel?

7. A resistance of 100 ohms is connected in parallel with another resistance of 120 ohms. These two are connected in series with a third resistor of 150 ohms. (a) What is the combined resistance of the three? (b) How much current will 200 volts deliver through the circuit? (c) How much current will the 100-ohm resistor get? (d) How much current will the 120-ohm resistor get? (e) What will be the drop in potential across the two resistors in parallel? (f) What will be the drop in potential across the 150-ohm resistor?

Chapter 9

Electrolysis

There are three classes of liquids:

1. Nonconductors (insulators), such as oils and turpentine.
2. Conductors, which pass electrical current without deterioration. Mercury and molten metals are examples.
3. Those liquids that will decompose when conducting an electrical current. Dilute acids, solutions of metallic salts, and certain fused solid compounds are of this kind.

Terminology

The liquids of class 3 are of interest because the electricity is transferred through these liquids by carriers that are called *ions*. The liquid is called the *electrolyte*. The transfer of ions is termed *electrolysis*. The apparatus used is termed the *electrolytic cell*. The plates immersed in the electrolyte, for the purpose of current entry and exit, are termed *electrodes*.

The electrode through which the current enters the electrolyte is called *cathode*. The electrode through which the current leaves the electrolyte is called the *anode*. See Figure 9-1, which illustrates the nomenclature used with an electrolytic cell.

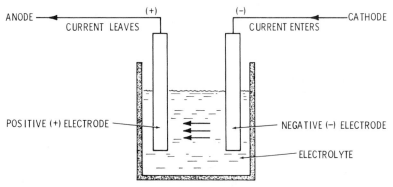

Figure 9-1   Electrolytic cell and terminology.

Chemistry of Electrolysis

The electrolyte will dissociate into positive (+) and negative (−) ions. The positive ions are attracted to the cathode and the negative ions are attracted to the anode. See Figure 9-2.

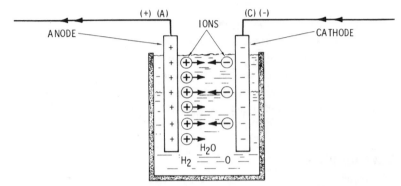

Figure 9-2   Conduction of hydrogen and oxygen ions in an electrolytic cell.

Water is usually used as the base for the electrolyte, with acids or metallic salts added to it. Hydrochloric acid (HCl) is often used as the additive to water (H_2O) as the base. The hydrochloric acid dissociates into positive hydrogen ions and negative chloride ions. This may be written in the equation

$$HCL = H^+ + CL^-$$

See Figure 9-3. Ionization in no manner changes the properties of the atoms. Hydrogen ions, such as H^+, will affect such chemical indicators as litmus paper, while hydrogen gas (H_2) won't affect litmus.

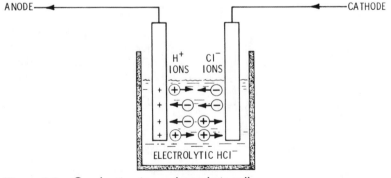

Figure 9-3   Conduction in an electrolytic cell.

Probably the simplest electrolysis might be the electrolysis of water (see Figure 9-4). Water is composed of two parts of hydrogen and one part of oxygen (H_2O). A jar is filled with water to which a little sulfuric acid (H_2SO_4) has been added. Two test tubes are filled with water and inverted into the electrolyte. An electrode of platinum is inserted into the down end of each test tube. These electrodes are connected to a battery, as illustrated in Figure 9-4. As the current passes through the water, it is broken down into two parts of hydrogen (H_2) in the right-hand test tube and into one part of oxygen (O) in the left-hand test tube. The right-hand test tube will have the water replaced in it twice as fast as in the left-hand tube, illustrating the one and two parts just mentioned.

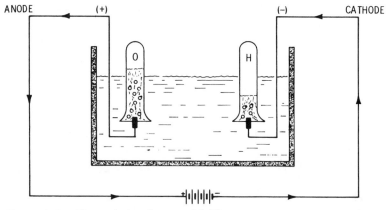

Figure 9-4   Electrolysis of water (H_2O).

The following will give you a better idea of the mechanics of the electrolysis of water. The letter e in the formulas will represent the negative electronic charge. As shown in Figure 9-2, the H^+ ions are attracted to the cathode. The sulfuric acid will also have H^+ ions and $SO_4^=$ ions. The H^+ ions drift to the cathode and the $SO_4^=$ ions, with two electrons, drift to the anode.

Each H^+ ion, upon reaching the cathode, combines with an electron there, forming a hydrogen atom, and the atoms combine to form a molecule and escape as hydrogen gas. The equation for this is

$$2H + 2e \rightarrow H_2\uparrow$$

in which the upward arrow represents the release of hydrogen into the test tube.

The $SO_4^=$ ions, when they reach the anode, give up their electrons and combine with water to form more H_2SO_4 and also liberate oxygen as a gas into the left-hand test tube. The formula for this is

$$2SO_4^= - 4e + 2H_2O \rightarrow \underbrace{4H^+ + 2SO_4^=}_{2H_2SO_4} + O_2\uparrow$$

Thus the water is being divided into two parts hydrogen and one part oxygen, as shown in Figure 9-4.

Electroplating

Just a word or two should be said about electroplating, since it is a form of electrolysis.

In electroplating, metallic ions are deposited on the cathode, plating it. The anode is composed of the same metal of which we wish the plating to be. Thus an anode of silver is used when silver plating and an anode of copper is used when copper plating.

The electrolyte varies with the plating being done. For copper plating, blue vitriol (copper sulfate) would be used.

Corrosion

Corrosion is the deterioration of metal by chemical or electrochemical reaction (electrolysis) with its environment. Moisture and oxygen are essential for corrosion. An acid or alkaline solution will accelerate the action.

The *National Electrical Code*, Section 344.10(B) requires that metallic conduit not be buried in earth or concrete unless additional corrosion protection is used. Metal conduits in earth will often corrode away, due to electrolysis caused by moisture, acids, and alkalis in the soil.

Table 9-1 gives the galvanic series (in part), detailing which metals are most affected and which are least affected by electrolysis. The upper end is least noble or anodic, and as you will recall, the anode in electroplating gives up its metal to the cathode.

The essentials for electrolysis are moisture, soil alkalinity or acidity, and the presence of two dissimilar metals. The least noble metal will be corroded away by the most noble metal.

In the days of DC-operated electric streetcars, the corrosion of water mains was a problem. One day the rails were made the cathode (–); this would corrode the water pipes. The next day the rails were made the anode (+) and this would make the pipe the cathode, thus putting back the metal removed.

Table 9-1   Galvanic Series of Metals

Corroded End—Least Noble (Anodic)

Potassium	Inconel (active)
Sodium	Brasses
Magnesium	Copper
Magnesium alloys	Bronzes
Zinc	Copper-Nickel alloys
Aluminum 2S	Monel
Cadmium	Silver Solder
Aluminum 17 ST	Nickel (passive)
Steel or iron	Inconel (passive)
Cast iron	Chromium-iron (passive)
Chromium-iron (active)	18-8 Stainless (passive)
Ni-resist	18-8-3 Stainless (passive)
18-8 Stainless (active)	Silver
18-8-3 Stainless (active)	Graphite
Lead-Tin Solders	Platinum
Lead	Gold
Tin	
Nickel (active)	
	Protected End—Most Noble (Cathodic)

The streetcar item was put in here for a point. Often AC is accused of causing corrosion of water pipes by using them for grounding purposes. This just can't be, as with 60-Hz current the direction of flow changes 120 times per second. Thus, one 1/120 of a second the grounding is a cathode, but the next 1/120 of a second it is an anode. Extensive tests have been run, and all the blame for corrosion from AC grounding has always come up as being without foundation.

An interesting case of corrosion is a galvanized (zinc-covered) hot-water tank in a home that corroded out in one year. The ground rod for the electric system was blamed. What really happened was that the heating system was hot water with a copper coil immersed in the hot-water system and then in turn going to the hot-water tank. Looking at Table 9-1, we find that copper is more noble than zinc. The answer was to put insulated couplings between the copper coil and the hot-water tank, or a sacrificial anode, such as magnesium, in the water tank.

Corrosion (electrolysis) is also the reason that aluminum shouldn't be directly connected to copper. The copper will, in plain words, eat the aluminum. This is electrolysis, and moisture and oxygen will hasten the problem.

Another problem of electrolysis is due to battery action, or local electrolysis set up by impurities in metals or coatings of metals buried in the earth. See Figure 9-5. The impurities in the zinc coatings or in reclaimed metals set up battery action, or local currents around the impurity, and electrolysis takes place, causing the pipe or other surface to wind up with holes in it.

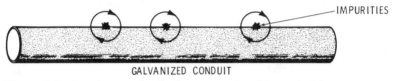

Figure 9-5   Local corrosion by battery action (electrolysis).

Questions

1. Which of the three classes of liquids are we concerned with in electrolysis?
2. What is the electrolyte?
3. What is the term for the transfer of ions?
4. What are the plates used for passing current into an electrolytic cell called?
5. Does the current enter or exit from the cathode of an electrolytic cell?
6. What two kinds of ions are associated with electrolysis?
7. Give the chemical formulas for electrolysis of water, using a weak solution of sulfuric acid in the electrolyte.
8. In your own words, explain the process of electroplating.
9. What is corrosion?
10. In the galvanic series, what does "most noble" indicate?
11. In the galvanic series, what does "cathodic" indicate?
12. Explain battery action, as it pertains to electrolysis.

Chapter 10

Primary and Secondary Cells

When two different metals are placed in contact with one another in air, one metal becomes positive and the other negative. This charge is very feeble.

The Voltaic Cell

Volta, a professor at the University of Pavia, experimented with this phenomenon. He took discs of zinc and copper, placed them in a pile alternately, and separated them with felt discs saturated with vinegar or other dilute acid. Such a stack is capable of giving a shock and will continue to do so as long as the felt discs are kept moist.

A small voltaic cell, which is named in honor of Volta, is shown in Figure 10-1. The zinc element or electrode, when immersed in the acid, starts to dissolve, leaving its electrons behind and the zinc goes into solution as Zn^{++} ions. This action stops very shortly, as the zinc plate becomes negative and ceases to throw off positive ions. The amount of zinc thus dissolved is very minute.

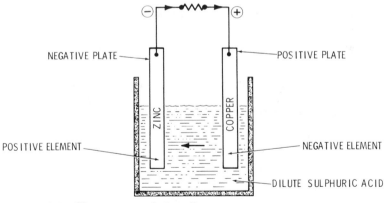

Figure 10-1   Elementary voltaic cell.

If the negative zinc plate is connected externally by a wire to the copper plate, an electrical current will flow from the negative zinc plate through the external wire to the copper plate (positive). As the negative charge of the zinc plate is thus removed, more Zn^{++} ions go into the solution; thus more electrons go on the zinc plate and the energy of the dissolution of the zinc into the acid is converted into electrical energy.

What about the copper plate? The same action occurs with the copper plate. Cu^{++} ions are given into the solution, leaving the copper plate negative also.

These two negative charges from the zinc and copper are unequal in quantity, as illustrated in Figure 10-2. The copper plate has a potential of 0.81 V and the zinc plate a potential of 1.86 V, both negative. Now $1.86 - 0.81 = 1.05$ V in favor of the zinc plate. So when the circuit is closed, as in Figure 10-1, the current flows from the zinc to the copper, through the resistor.

Actually, one may say the difference of potential between the plates is a measure of the difference of their tendencies to oxidize.

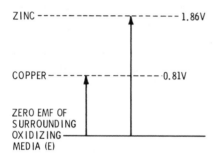

Figure 10-2 Difference in potential of the plates.

You have noticed that this apparatus was called a voltaic cell. One such unit is a *cell*. If more than one is connected in series or parallel, the combination becomes a *battery*. See Figure 10-3.

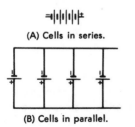

(A) Cells in series.

(B) Cells in parallel.

Figure 10-3 A battery.

Primary Cells

In a cell there must be two metals that differ in that one oxidizes more readily than the other. Note that in Figure 10-1 the zinc is called the positive element and the copper is called the negative element. Now refer to Table 9-1, "Galvanic Series of Metals," and you will find that zinc is anodic compared to copper, which is more cathodic than zinc.

Impure zinc will dissolve in sulfuric acid, while pure zinc won't. In the voltaic cell in Figure 10-1, the zinc will dissolve when the external circuit is closed. The chemical formula is

$$H_2SO_4 \rightleftharpoons 2H^= + SO_4^=; \text{ then } Zn^{++} + SO_4 \rightarrow ZnSO_4$$

The hydrogen is given up at the copper plate. The zinc is dissolved in proportion to the current flowing from the cell. This is the *primary cell* and is considered to be unrechargeable. It is replenished by replacing the zinc plate and the electrolyte.

Commercial zinc usually has impurities in it, such as iron, arsenic, or some other metal. These particles are small, but refer to Figure 10-4 for the local battery action that transpires, causing the zinc to waste away. The impurity, zinc, and acid form a small cell that is shorted, so current flows. This local action may be stopped by amalgamating the zinc by rubbing it with mercury. The amalgamation loosens the impurities and they float to the surface. Mercury is sometimes added to the zinc while it is in a molten mass.

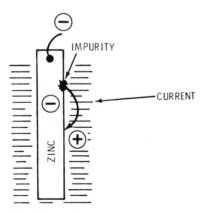

Figure 10-4 Local battery action.

Some of the liberated hydrogen clings to the positive plate and thus effectively reduces the effective surface of this plate. This reduction reduces the effective emf from the cell. This action is known as *polarization.*

Depolarizers are added to the cell to stop polarization. Depolarizers may be liquids or solids. Some liquid depolarizers are

Nitric acid

Chromic acid

Bichromate of potash

Bichromate of soda

Nitrate of potash

Some solid depolarizers are

Black oxide of manganese

Oxide of copper

Peroxide of lead

Oxide of lead

All depolarizers abound in oxygen. The current releases this oxygen, which combines with the hydrogen that caused the polarization.

A few primary cells will be discussed in brief, so that you may gain further knowledge of them. As stated before, the primary cell is replenished by replacing the negative plate, if it has been depleted, and replacing the electrolyte.

A *secondary cell* is one that is rechargeable, without replacing the plate or electrolyte. Of course, there is an end to this, as we have seen when we have had to replace an automobile battery, which is a rechargeable secondary-cell combination.

The *Daniell cell,* illustrated in Figure 10-5, uses the electrochemical method of avoiding polarization. The Daniell cell is based on the theory that whenever a current passes from a metal to a solution of its own salt, metallic atoms are dissolved into the solution.

The zinc plate is placed in a dilute solution of zinc sulfate ($ZnSO_4$) with a little sulfuric acid (H_2SO_4). The copper plate is placed in a porous, unglazed earthenware cup. This will pass ions, but not the solution. The solution in the cup is copper sulfate (blue vitriol, $CuSO_4$), with some crystals of $CuSO_4$ placed in the cup containing the copper plate.

The zinc dissolves, leaving electrons on the zinc plate, and gives off Zn^{++} ions. These ions combine with the $SO_4^=$ ions to form zinc sulfate.

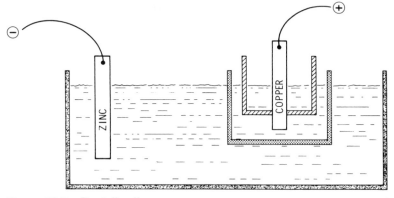

Figure 10-5   Daniell cell.

$$Zn \rightarrow Zn^{++} + 2e \quad \text{and} \quad Zn^{++} + SO_4^{=} \rightarrow ZnSO_4$$

The Cu^{++} ions of the depolarizer take electrons from the copper plate, leaving it positive, and deposit metallic copper on the positive plate:

$$CuSO_4 \rightleftharpoons Cu^{++} + SO_4^{=}$$
$$Cu^{+} + 2e \rightarrow Cu \downarrow$$

The hydrogen ions from the sulfuric acid (H_2SO_4) combine with the $SO_4^{=}$ ions of the depolarizer and make sulfuric acid:

$$2H^{+} + SO_4^{=} \rightarrow H_2SO_4$$

During the use of the cell, the zinc wastes away while the copper gains weight.

The *gravity cell* has the same action as the Daniell cell, but zinc sulfate is lighter than copper sulfate, so the two keep separated by specific gravity. See Figure 10-6. The gravity cells were used extensively in telegraphy.

The *Leclanche cell* uses zinc and carbon plates. The solution contains sal ammoniac, which is ammonia chloride (NH_4Cl). See Figure 10-7. The depolarizer used is manganese oxide (MnO_2). This cell supplies about 1.5 V emf. The accompanying formulas are

$$Zn^{++} + 2Cl^{-} \rightarrow ZnCl_2 \text{ (near negative plate)}$$
$$2NH_4^{+} + 2e + MnO_2 + H_2O \rightarrow 2NH_4OH + MnO$$

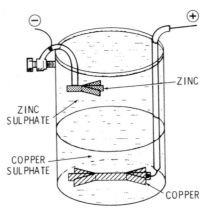

Figure 10-6   Gravity cell.

Dry cells are in reality a form of the Leclanche cell. The outer case is composed of zinc, which is the negative electrode. The center post is carbon and is the positive post (electrode). The chemicals are manganese dioxide (depolarizer), ground coke, sal ammoniac, and zinc chloride. The zinc chloride is added to lengthen the life of the cell by retarding local action.

There are other primary cells, but since the basic theories have been discussed, no further subject matter will be covered.

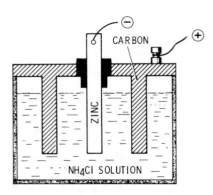

Figure 10-7   Leclanche cell.

Secondary Cells

If a charging current is sent through a Daniell cell in reverse direction to the normal flow, this will dissolve some of the copper electrode into the sulfuric acid, making copper sulfate. At the same time some of the zinc atoms will be taken out of the zinc sulfate solution and will be deposited on the zinc electrode. The Daniell cell may be replenished by recharging, instead of by replenishing the zinc plate and electrolyte.

A *storage* or *secondary cell* is rechargeable, instead of needing to have the chemicals or the like replenished. We associate the secondary cell or secondary or storage battery with the automobile battery.

The automobile battery is a lead and sulfuric acid battery. It is made up of two lead electrodes immersed in a solution of sulfuric acid (H_2SO_4) and water (H_2O). The plates are connected to a battery charger as shown in Figure 10-8.

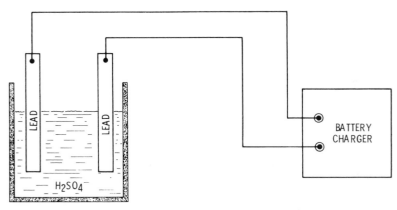

Figure 10-8   Charging a lead-acid cell.

A charged storage cell consists of a positive plate of lead oxide (PbO_2), a negative plate of spongy lead (Pb), and sulfuric acid (H_2SO_4). See Figure 10-9. The positive plate is dark brown and the negative plate is light gray.

As the cell discharges, both plates have the active materials turned into lead sulfate ($PbSO_4$) (see Figure 10-10), and it is this condition in which they appear when it becomes necessary to recharge the cells.

The chemical symbols for lead and some of its compounds are as follows:

Lead	Pb
Red lead	Pb_3O_4
Peroxide of lead or lead dioxide	PbO_2
Monoxide of lead	PbO
Sulfate of lead	$PbSO_4$
Hard sulfate of lead	Pb_2SO_5

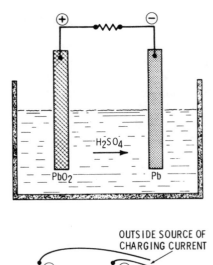

Figure 10-9   Charged storage cell.

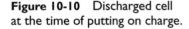

OUTSIDE SOURCE OF
CHARGING CURRENT

Figure 10-10   Discharged cell at the time of putting on charge.

From previous information, we found that H_2SO_4, when charging or discharging a cell, dissociates into hydrogen H^+ ions and sulfate $SO_4^=$ ions. The lead dioxide (PbO_2) in Figure 10-9 reduces to lead monoxide (PbO). This occurs in the positive plate.

The equations associated with the discharge as are follows:

Negative Plate:

$$Pb \rightarrow Pb^{++} + 2e$$
$$Pb^= + SO_4^= \rightarrow PbSO_4$$

Positive Plate:

$$PbO_2 + 2H^+ + 2e \rightarrow PbO + H_2O$$

PbO_2 is inactive with H_2SO_4, but PbO, which forms at the positive plate when reduced from PbO_2, reacts readily with H_2SO_4. Thus,

$$PbO + H_2SO_4 \rightarrow PbSO_4 + H_2O$$

So, on discharge, both plates are coated with $PbSO_4$, and the H_2SO_4 is partially converted to water (H_2O).

The voltage of each cell starts at 2.1 V and drops to 2.0 V, where it remains throughout the discharging cycle until toward the end, and then it drops off rapidly.

During the charging from a charging source, the $PbSO_4$ on the negative plate is restored to lead (Pb) and the positive plate is restored to lead oxide (PbO_2) (see Figure 10-9); and the H_2SO_4, which was diluted with H_2O, reverts to its original concentration. Thus,

$$\textit{Negative Plate: } PbSO_4 + 2H^+ + 2e \rightarrow Pb + H_2SO_4$$
$$\textit{Positive Plate: } PbSO_4 + SO_4^= - 2e \rightarrow Pb(SO_4)_2$$

The $Pb(SO_4)_2$ is a plumbic sulfate that is unstable in the presence of water and will break down:

$$Pb(SO_4)_2 + H_2O \rightarrow PbO_2 + 2H_2SO_4$$

The positive plate is restored upon charging to the state shown in Figure 10-9.

It is not practical to form plates from scratch, so in manufacturing lead-acid cells, grids of lead are formed and the proper compounds are pressed into the grids.

The state of charge is measured by a *hydrometer*, which checks the specific gravity of the electrolyte and tells us the state of charge. A hydrometer is a weighted glass tube, marked in specific gravity, which is placed inside of a larger glass tube into which the electrolyte is pulled. The weighted tube seeks its proper flotation level.

A discharged lead-acid cell has a low specific gravity, while the low specific gravity of a charged cell is around 1.275. This basically tells us whether the electrolyte is highly acidic or highly watered. Table 10-1 shows the specific gravity of the electrolyte in terms of so many parts water to one part acid.

A discharged battery is very apt to freeze, and Table 10-2 will tell us why.

Another type of storage cell is the Edison storage cell or nickel-alkaline storage cell, conceived by Thomas A. Edison. The positive

Table 10-1   Sulfur-Acid Battery Solution

Specific Gravity	Parts Water to 1 Part Acid
1.200	4.4
1.225	3.7
1.250	3.2
1.275	2.8
1.300	2.5

Table 10-2   Freezing Temperature of Acid Batteries

Specific Gravity	Degrees Fahrenheit	Specific Gravity	Degrees Fahrenheit
1.275	−85	1.125	+13
1.250	−62	1.100	+19
1.225	−16	1.050	+25
1.175	−4	1.000	+32
1.150	+5		

electrode is nickel oxide and the negative electrode is finely divided iron. The electrolyte is potassium hydroxide. The following symbols will be useful in showing the chemical reactions:

KOH	Potassium hydroxide
K	Potassium
OH	Hydroxide
Fe	Iron
Ni	Nickel
NiO	Nickel oxide
NiO_2	Nickel dioxide

The following are the chemical formulas involved:

Discharge

Negative Plate: $\quad Fe + 2OH^- - 2e \rightarrow FeO + H_2O$

Positive Plate: $\quad NiO_2 + 2K^= + H_2O + 2e \rightarrow NiO + 2KOH$

Charge

Negative Plate: $FeO + 2K^= + H_2O + 2e \rightarrow Fe + 2KOH$

Positive Plate: $NiO + 2OH^- - 2e \rightarrow NiO_2 + H_2O$

The information given in this chapter gives the basic theory of what takes place chemically in both primary and secondary cells.

Questions

  1. Describe a voltaic cell.

  2. What is the voltage of a voltaic cell, and why?

  3. What is a cell?

  4. What is a battery?

  5. Give the chemical formulas for a voltaic cell.

  6. Explain depolarizers and their purpose. List some depolarizers.

  7. Explain the operation of a Daniell cell.

  8. Which plate dissolves in a Daniell cell?

  9. Give the chemical formulas for a Daniell cell.

10. What is a gravity cell?

11. How does a gravity cell differ from a Daniell cell?

12. Describe a Leclanche cell.

13. What is a primary cell?

14. What is a secondary cell?

15. Explain the construction of the common dry cell.

16. What is our most common secondary cell?

17. Of what are the plates of a lead-acid cell composed when discharged?

18. Of what are the plates of a lead-acid cell composed when charged?

19. Explain fully the processes in charging and discharging a lead-acid cell.

20. What is a hydrometer, and for what is it used?

21. Explain the operation of an Edison cell.

22. Give formulas for the discharge of an Edison cell.

23. Give formulas for the charging of an Edison cell.

Chapter 11

Electromagnetism

Hans C. Oersted (1777–1851), a Danish physicist at the University of Copenhagen, observed in 1819 that a compass needle was affected by a voltaic pile. Upon further experimenting, he discovered that a compass needle that was placed immediately above or below a conductor carrying current, as in Figure 11-1, was deflected.

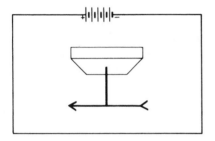

Figure 11-1 Compass needle is deflected by a current through a conductor.

This effect may be further demonstrated by passing a conductor through a paper on which iron filings have been sprinkled. When direct current is passed through the conductor, the iron filings will arrange in a configuration such as in Figure 11-2.

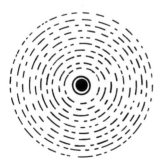

Figure 11-2 Iron filings around a current-carrying conductor.

In Figure 11-3, the lines of the magnetic field around a conductor are more clearly illustrated. The lines of force around a current-carrying conductor form in circular paths. In order to establish in which direction those lines of force travel, we will use Figures 11-4 and 11-5. In Figure 11-4, the current in the conductor is going

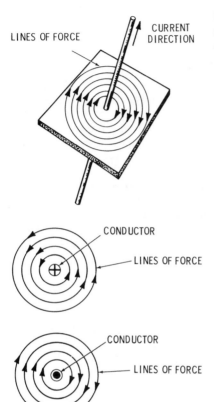

LINES OF FORCE

CURRENT
DIRECTION

Figure 11-3 Magnetic field around a current-carrying conductor.

CONDUCTOR

LINES OF FORCE

Figure 11-4 Magnetic field of a current moving into the page.

CONDUCTOR

LINES OF FORCE

Figure 11-5 Magnetic field of a current moving out of the page.

away from us and is represented by the end of the conductor having a plus in it; the lines of force or flux go around counterclockwise. In Figure 11-5, the current is coming toward us as represented by the dot in the circle; here the lines of flux are clockwise.

An easy way to remember is by means of the *left-hand rule*. In your imagination, place your left hand around a conductor, as in Figure 11-6, so that your thumb points in the direction of the current (negative to positive). Then your fingers will be pointing in the direction of the lines of force.

Refer to Figure 11-7 to determine the direction in which a compass needle will be deflected.

Parallel conductors carrying currents in the same direction attract each other. See Figure 11-8. Conductors *A* and *B* are carrying current away from us, so the lines of force are counterclockwise.

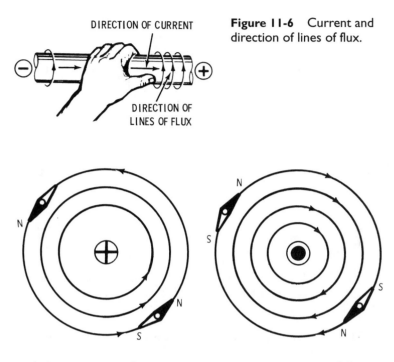

DIRECTION OF CURRENT

DIRECTION OF
LINES OF FLUX

Figure 11-6 Current and direction of lines of flux.

(A) Current going into the page.

(B) Current coming out of the page.

Figure 11-7 Direction of deflection of a compass needle.

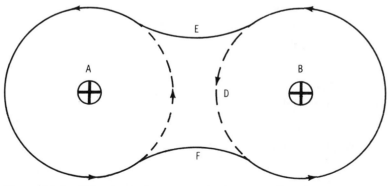

Figure 11-8 Wires carrying current in the same direction attract each other.

Instead of circling A and B separately, as at C and D, they combine and encircle both conductors, as at E and F.

Wires carrying current in opposite directions repel each other. See Figure 11-9. Conductor A is carrying current away from us and conductor B is carrying current to us. Lines of force around A are counterclockwise and clockwise around B. Since the lines of force are oriented in opposite directions, they won't combine as in Figure 11-8, but will tend to push the wires apart.

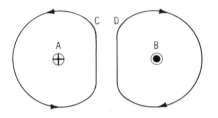

Figure 11-9 Wires carrying current in opposite directions repel each other.

Maxwell's rule states that every electrical circuit is acted upon by a force that urges it in such a direction as to cause it to include within its embrace the greatest possible number of lines of force.

In explanation of Maxwell's rule we shall use Figure 11-10. In Figure 11-10A, there is a circuit doubled back on itself. The lines of

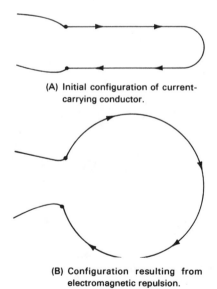

(A) Initial configuration of current-carrying conductor.

(B) Configuration resulting from electromagnetic repulsion.

Figure 11-10 Illustration of Maxwell's rule.

force around half the wire will oppose the lines of force around the other half, as was covered in the explanation of Figure 11-9. This means the wire will have a tendency to be pushed apart and, if free to move, would theoretically take the shape of the circuit shown in Figure 11-10B, which would be a circle.

A paraphrase of Maxwell's rule is that every electrical circuit tends to so alter its shape as to make the magnetic flux through it a maximum. In this paraphrase, you have the answer that explains motor action and also the action of many measuring instruments.

In explanation, every electric motor has a loop of wire that carries current. This loop is placed in such a position in a magnetic field that the lines of force pass parallel to, but not through, it. From Maxwell's rule, the loop tends to turn in such a direction so as to include within it the lines of force of the magnetic field. It is suggested that this action be reviewed and remembered, as it will have a far-reaching effect in later chapters.

One of these far-reaching effects will be covered under a chapter concerned with fault currents, but it will be good to touch on this matter now, while we are involved in one important effect caused by fault currents.

On large-capacitance circuits, which will have high currents available should the conductors of a circuit short together, Maxwell's rule must be prepared for before the short occurs. When the short occurs, the magnetic forces tending to cause the circuit to embrace the greatest possible number of lines of force will tend to throw the conductors apart. These magnetic stresses become very great. In switchgear, for instance, where bus bars are the conductors, it becomes an engineering problem to design the bus bars so that they won't be torn from their mountings. Therefore, they have to be rigidly supported and bolted into place.

The same must be done with cable trays. Here the conductors should be tied down securely so that they won't be thrown out of the tray or destroy the tray.

In Figure 11-11, the lines of force from the magnet go from N to S and the current in the conductor is coming toward us, so the flux

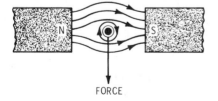

FORCE

Figure 11-11 Composite of a magnetic field and a current-carrying conductor.

around the conductor is clockwise. One might compare these lines of force to rubber bands. Thus the lines of force of the magnetic poles tend to straighten out and push the conductor down.

The *right-hand rule* is illustrated in Figure 11-12. The right hand is cupped over the pole piece as shown; the thumb represents the direction of motion, the index finger the direction of the flux, and the middle finger the direction of the current.

Figure 11-12 Right-hand rule.

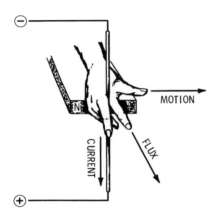

Galvanoscope

Figure 11-1 showed a magnetic needle under a wire carrying current, as did Figure 11-7. A *galvanoscope* is such a device. Figure 11-13 will be used in the explanation. Figure 11-13A shows a simple galvanoscope, similar to the one previously illustrated. Current will deflect the magnetic needle, which is suspended by a thread. If the current is very feeble, the deflection will be hard to notice. To overcome this, more turns are added, as in Figure 11-13B. Thus, if there are 100 turns, the effect will be 100 times as much as the effect of

(A) Simple galvanoscope.   (B) Multiplying effect of many turns.

Figure 11-13 The galvanoscope.

the simple galvanoscope. The theory discussed here will be applied later in galvanometers and electrical measuring instruments.

Solenoids

In Figure 11-8 the effect of conductors in parallel is shown. Notice that conductors in parallel, with the currents in the same direction, cause the lines of force to embrace the conductors as one. In solenoids and electromagnets, this phenomenon is taken advantage of to increase the strength of the solenoids and magnets.

Figure 11-14 is used to illustrate. Here there is a soft iron core, A, wound with turns of wire. As you see, the turns are parallel and the current flows in the same direction in all turns. Notice that the lines of force are all in the same direction on either side, so there is an addition of the lines of force in proportion to the number of turns. The current has to be in the same direction in each turn as the turns are in series. The lines of force add up and concentrate at poles B and C of the iron core.

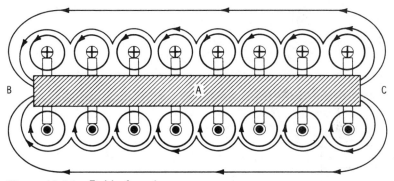

Figure 11-14   Field of an electromagnet.

It will be mentioned here that a circuit doubled back on itself produces no magnetic field. This is because the magnetic field of one wire cancels out the effect of the magnetic field of the other wire, since the fields are equal and opposite.

Mechanical functions are performed by solenoids such as the one illustrated in Figure 11-15. The coil M is wound on a nonmagnetic form. Plunger A is drawn into coil M when the coil is energized, pulling lever L up, actuating some mechanical function. When coil M is deenergized, the spring S pulls the lever back down.

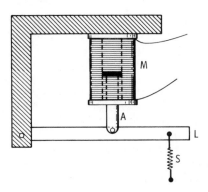

Figure 11-15   Principle of a solenoid.

Solenoids and electromagnets are used extensively in the electrical industry. Figure 11-16 illustrates the principle of a relay. The electromagnet M is energized, pulling armature A to it, and the contact on the armature A makes contact with the contact C, closing a circuit between 1 and 2. When the magnet is deenergized, the spring S pulls the armature A away, opening the contact between A and C.

Some circuit breakers are triggered by relays when overcurrents occur.

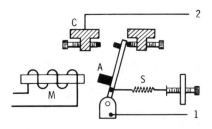

Figure 11-16   Principle of a relay.

Questions

1. Explain the effect that current through a conductor will have on a compass needle.

2. Show the symbols used on conductor ends to indicate the direction of current.

3. Fully describe the left-hand rule for the direction of magnetic flux around a current-carrying conductor.

4. What happens to magnetic lines of force produced by conductors with current in both conductors in the same direction?

5. What is the effect of magnetic lines of force around two conductors when the currents in the two conductors are flowing in opposite directions?

6. Explain Maxwell's rule as it concerns the magnetic effect of an electrical current in a circuit.

7. What is a galvanoscope?

8. How may the magnetic effect of current be magnified in a galvanoscope?

9. Define a solenoid. Sketch and explain.

10. Define an electromagnet. Sketch and explain.

Chapter 12

Laws Governing Magnetic Circuits

Since certain definitions are very important to this discussion, the following definitions will be given here:

> *Maxwell* (Mx): Unit of magnetic flux, one magnetic line of force.
>
> *Gauss* (G): Unit of flux density *B*, equal to one maxwell per square centimeter.
>
> *Gilbert* (Gb): Unit of magnetomotive force (mmf, the force by which a magnetic field is produced), equals $10/4\pi$ ampere-turns (At).
>
> *Oersted* (Oe): Unit of magnetic field strength or magnetizing force equal to $1000/4\pi$ ampere-turns per meter.
>
> *Permeability* (μ: Greek letter "mu"): Expresses the relationship of flux density produced in a magnetic substance to the field intensity that occasions it.

At this time we will be using the above terms, plus some additional definitions, in order that we might gain knowledge concerning magnetic calculations. There are definite similarities between magnetic circuits and electrical circuits. These should be noted, as they will make the subject easier to understand.

Figure 12-1 will be used to illustrate the gauss. As noted, the end is 2 cm by 2 cm, or 4 cm^2. Let the number of lines of force through 1 cm^2 be 10,000; the flux density is then said to be 10,000 gausses,

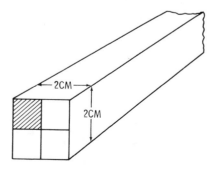

Figure 12-1 Representation of a gauss.

or 10 kilogausses. If now the flux density B is multiplied by the cross-section s, the total flux ϕ in maxwells will be

$$\phi = B \times s = 10,000 \times 4 = 40,000 \text{ lines of force}$$

The electrical force that produces the magnetic lines of force is called the *magnetizing force*. This force is usually produced by a coil of wire that carries an electrical current. The magnetizing force is represented by H. One unit of magnetizing force H will produce one magnetic line of force B per square centimeter in air.

Permeability (μ) is the ratio of magnetic flux density B to the magnetizing force H. So,

$$\text{Permeability} = \frac{\text{Magnetic Flux Density}}{\text{Magnetizing Force}} \quad or \quad \mu = \frac{B}{H}$$

The permeability of air is 1, since in air B always equals H, so $B/H = 1$.

If a current from a battery is passed through an air-core solenoid as in Figure 12-2, the magnetic lines of force will affect the suspended magnetic needle N. Now, if we provide an iron core for the same solenoid in Figure 12-2, we have iron instead of air for the core. With the same current through the solenoid, the magnetic needle N will be deflected more, although the same magnetizing force was applied to the solenoid. The reason for the greater deflection when the iron core is added is the result of a much increased magnetic flux density B, with the addition of the iron core.

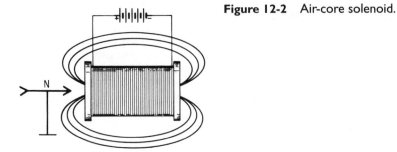

Figure 12-2 Air-core solenoid.

In explanation, suppose that with the air core, the magnetizing force of 50 units of H produced a flux density B of 50 lines of force per square centimeter. The permeability would be $B/H = \mu = 50/50 = 1$, which is always true for air. After the addition of the iron core, we still have 50 units of magnetizing force H, but now have, with 15,000 lines of force per square centimeter,

$$\mu = \frac{B}{H} = \frac{15,000}{50} = 300$$

as the permeability of the iron core that we inserted into the solenoid. This indicates that the iron core conducted 300 times the number of magnetic lines of force as the air core.

Table 12-1 will give you a better idea of the variations in permeability of different substances.

Table 12-1   Permeability of Magnetic Substances

	Mmax (Gauss/Oersted)	B (Gauss)
Cobalt	170	3000
Iron-cobalt alloy (Co 34%)	13,000	8000
Iron, purest commercial annealed	6000 to 8000	6000
Nickel	400 to 1000	1000 to 3000
Pennalloy (Ni 78.5%, Fe 21.5%)	Over 8000	5000
Perminvar (Ni 45%, Fe 30%, Co 25%)	2000	4
Silicon steel (Si 4%)	5000 to 10,000	6000 to 8000
Steel, cast	1500	7000
Steel, open-hearth	2000 to 7000	6000

Strength of a Magnetic Pole

A unit magnetic pole may be considered to be a point that sends out enough lines of force to produce a flux density of one magnetic line to every square centimeter of a spherical surface situated 1 cm from the pole and centered at the pole point. There will be as many lines of force concentrated at the pole point as there are square centimeters on the surface of a sphere 1 cm in radius. A sphere of 1 cm radius has a surface area of 4π cm^2 (12.57 cm^2). Therefore, every magnetic pole of unit strength has 4π lines of force emanating from it or entering into it.

Example
A magnet with a strength of 15 unit poles will have $15 \times 4\pi$ or 189 lines flowing out of the north pole and into the south pole.

Intensity of Magnetizing Force

We have learned that the magnetizing force of an electromagnet is produced by current flowing through a coil of wire. The intensity of this magnetizing force per centimeter of length is expressed as follows:

$$H = \frac{4\pi \ IN}{10 \ l} = \frac{1.257 \ IN}{l}$$

where

H = intensity of magnetizing force per unit of length

I = current in amperes

N = number of turns in the coil

l = length of solenoid in centimeters

10 = constant to reduce amperes to absolute units

Magnetic Reluctance

There is resistance to the flow of magnetic lines, called *reluctance,* the symbol for which is R. No unit term is currently used for reluctance.

The calculation of reluctance is not quite as simple as the calculations of ohmic resistance for electrical circuits. This is due to the peculiar tendency of magnetic substances to reach a saturation point. This indicates that the permeability of a substance is not a fixed quantity, but changes with flux density.

The permeability of a piece of cast iron with a flux density of 4000 lines per square centimeter is 800. It is found that if the flux density is increased to 5000 lines, the permeability will fall to 500.

There are tables available in handbooks for permeabilities of various magnetic substances, under different flux densities. To illustrate our point, Table 12-2 covers one type of wrought iron.

Table 12-2   Flux Density, Magnetic Force, and Permeability

B	H	μ	B	H	μ
1000	0.48	2080	9000	2.95	3050
2000	0.61	3280	10,000	4.32	2310
3000	0.78	3850	11,000	6.70	1640
4000	0.92	4340	11,500	9.46	1220

Table 12-2   (continued)

B	H	μ	B	H	μ
5000	1.08	4620	12,000	12.40	953
6000	1.20	5000	12,500	16.00	781
7000	1.40	5000	13,000	23.80	546
8000	2.00	4000			

The formula for magnetic reluctance is as follows:

$$R = \frac{l}{s\mu}$$

where

R = reluctance

l = length of magnetic circuit in centimeters

μ = permeability

s = cross-section of magnetic circuit in square centimeters

This formula indicates that reluctance increases directly with the length of the magnetic circuit and decreases as the product of the permeability and cross-section.

The resistance of a wire increases with the length and decreases inversely with its cross-sectional area and conductivity.

Ohm's law for an electrical circuit is $I = E/R$. Rowland's law for magnetic circuits is ϕ = mmf/R, or

$$\text{Magnetic Flux} = \frac{\text{Magnetomotive Force}}{\text{Magnetic Reluctance}}$$

The detailed formula for total flux is ϕ = mmf/R, or

$$\phi = \frac{4\pi\,(IN/10)}{l/\mu s} = \frac{1.257\,IN}{l/\mu s} = \frac{1.257\,IN\,\mu s}{l}$$

so that

$$\phi l = 1.257\,IN/\mu s$$

where

ϕ = total magnetic flux in maxwells

IN = ampere-turns

l = length of magnetic circuit in centimeters

s = cross-section of magnetic circuit in square centimeters

μ = permeability

1.257 = constant for centimeter measurements

mmf = magnetomotive force in gilberts

Thus, the magnetizing force is proportional to the ampere-turns. To further compare electric and magnetic circuits:

$$\textit{Electric Circuit} \quad 1 \text{ ampere} = \frac{1 \text{ volt}}{1 \text{ ohm}}$$

$$\textit{Magnetic Circuit} \quad 1 \text{ maxwell} = \frac{1 \text{ gilbert}}{1 \text{ unit reluctance}}$$

Since the magnetizing force is proportional to the ampere-turns (IN), 1 ampere through 10 turns will equal 10 ampere-turns. The same magnetizing force will be obtained with $\frac{1}{10}$ ampere and 100 turns, or 10 ampere-turns.

It is often required to find the ampere-turns (IN) necessary to produce a given total flux:

$$IN = \frac{\phi(l/\mu s)}{3.192} = \frac{\phi l}{3.192 \ \mu s}$$

where

IN = ampere-turns

ϕ = total magnetic flux in maxwells

l = length of magnetic circuit in inches

s = cross-sectional area of magnetic circuit in inches

μ = permeability

3.192 = constant for inch measurement

The difference between the intensity of magnetizing force (H) and the total magnetomotive force (mmf) may be best illustrated by the following formula:

$$H = \frac{1.257 \ IN}{l}$$

where

IN = ampere-turns

H = magnetizing force per centimeter of length

l = length of magnetic circuit in centimeters

The difference may be illustrated in Figure 12-3. The following are the meanings of the symbols used in the figure:

H = intensity of the magnetizing force per centimeter of length

F = total mmf for whole bar expressed in gilberts

B = flux density in lines of force per square centimeters, expressed in gausses

ϕ = total magnetic flux for the entire 4 cm², expressed in maxwells

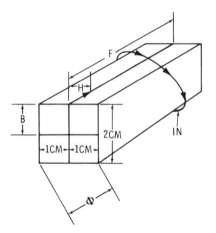

Figure 12-3 Application of magnetic terms.

Thus, IN may be expressed as

$IN = \phi l / 1.257 \; \mu s$

$IN = Bl / 1.257 \; \mu$

A practical application is as follows: It is desired to produce a flux ϕ of 20,000 maxwells in an iron ring (Figure 12-4) having a

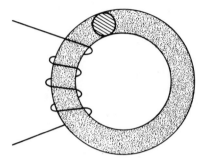

Figure 12-4   Iron ring and *IN*.

mean circumference of 150 cm and a cross-sectional area of 4 cm². Find the ampere-turns required. (Use a permeability of 3000.)

$$IN = \frac{\phi l}{1.257 \; \mu s} = \frac{20,000 \times 150}{1.257 \times 3000 \times 4} = 159$$

We could use 159 turns at 1 ampere or 10 turns at 15.9 amperes; the results would be the same.

Problem
Produce 65,000 maxwells in a ring with a mean circumference of 50 cm and 5 cm² in cross-section. Use 1.083 for the permeability.

$$B = \phi/s = \frac{65,000}{5} = 13,000 \; \text{lines/cm}^2$$

$$IN = \frac{65,000 \times 5}{1.257 \times 1083 \times 5} = 470 \; \text{ampere-turns}$$

Now, if the ring shown in Figure 12-4 was cut at the bottom (Figure 12-5) and the poles pulled apart 1 cm, this would increase the reluctance because of the added air gap. We would now use the formula

$$IN = \frac{\phi}{1.257} \left[\frac{l}{\mu s} + \frac{l^1}{\mu^1 s^1} \right]$$

where

> l^1 = length of air gap in centimeters
> μ^1 = permeability of air gap = 1
> s^1 = cross-section of air gap in square centimeters

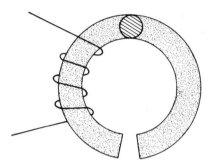

Figure 12-5 Iron ring with air gap.

$$IN = \frac{65,000}{1.257} \left(\frac{50}{1083 \times 5} + \frac{1}{1 \times 5} \right) = 10,370 \text{ ampere-turns}$$

Compare IN with the previous problem and we find that with the air gap, we now require 21 times the ampere-turns for the same flux.

Generators always have air gaps, so we should observe that the greater number of turns in a generator is because of air gaps.

Saturation of magnetic substances was discussed previously. Figure 12-6 shows a typical permeability curve. Note that point X (knee of curve) is about the most efficient point of operation, because additional ampere-turns don't materially affect the magnetic flux density B as compared to the extra magnetizing force H that is required.

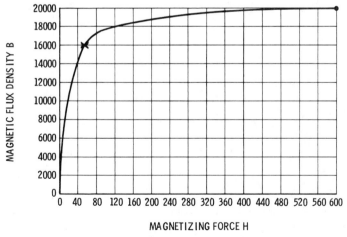

Figure 12-6 Typical permeability curve.

There is always a lagging of flux density or magnetization behind the application of the magnetizing force. This is termed *hysteresis*. Hysteresis is a loss, and the area within the hysteresis loops represents this loss (Figure 12-7).

Residual magnetism is always involved. If a piece of iron is subjected to an increase in magnetizing force and then this force is decreased to zero, there will be some retention of magnetism, which is termed *residual magnetism*.

Referring to Figure 12-7, we start at set zero and increase the magnetizing force in direction $+H$ with the flux density increasing in the direction $+B$ until point A is reached. Then, if the magnetizing force H is decreased, the flux doesn't follow the initial ascending

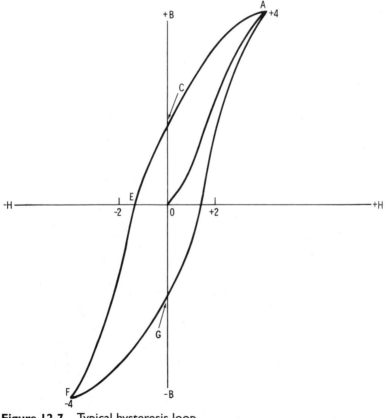

Figure 12-7   Typical hysteresis loop.

curve oA. It descends until H is zero, where the curve is at C. This is due to the retained magnetism, and oC represents this retained magnetism. All of this varies with the magnetic properties of the material involved.

If the current is reversed, the point E gives the magnetizing force required to bring the residual magnetism to zero. At this point the magnetism is reversed and becomes maximum at point F.

Formulas

$$B \times s = \phi$$

$$\mu = B/H$$

$$H = \frac{4\pi \; IN}{10l} = \frac{1.25 \; IN}{l}$$

$$R = \frac{l}{s \times \mu}$$

$$\phi = \text{mmf}/R$$

$$\phi = \frac{4\pi \times (I/10) \; N}{l \, (\mu \times s)}$$

$$\phi = \frac{1.257 \; IN \times \mu s}{l}$$

$$IN = \frac{\phi \times l}{1.257 \; \mu s} \quad \text{(metric)}$$

$$IN = \frac{\phi \times l}{3.129 \; \mu s} \quad \text{(for inches)}$$

Questions

1. Define a gauss.
2. Define a maxwell.
3. Define an oersted.
4. Define a gilbert.
5. Define permeability.

6. What is the permeability of air?
7. What is the permeability of silicon steel?
8. What is magnetic reluctance?
9. Give the formula for magnetizing force.
10. Give two formulas for Rowland's law.
11. Give a formula for ampere-turns.
12. Define and explain residual magnetism.
13. Define and explain hysteresis.
14. For a permeability curve, give your version of the "knee of the curve."

Chapter 13

Work, Power, Energy, Torque, and Efficiency

Energy is the ability to do work. Energy can neither be created nor destroyed. It may, however, be converted from one form to another. The conversion of one form of energy to another is accompanied by some form of loss. Potential energy is stored energy, such as in a battery. Kinetic energy is energy in motion.

A good example of energy is steam under pressure in a boiler. This is potential energy; there it is ready to go to work. By the same token, the energy in the steam was converted from burning of some type of fuel.

Another example is the spring of a watch that has been wound up and is ready to expend its energy into the work of running the watch. Still another form of energy is present in a car battery, ready to be released by chemical action.

When energy is released, it is capable of doing work. *Work* is the overcoming of opposition through a certain distance.

The amount of energy possessed is equal to the amount of work that the energy is capable of doing, less some losses.

Power is the rate of doing work. The faster work is done, the greater the power that will be required to do it.

It has been stated that energy is the ability to do work. Work is accomplished upon the release of energy. Work is the product of the moving force times the distance through which the force acts in overcoming opposition. Work may be measured in foot-pounds (ft-lb). One foot-pound is the work required to lift 1 lb a distance of 1 ft. Please note that no reference has been made to time.

Figure 13-1 illustrates work. A man pulls a 50-lb weight up 8 ft and therefore does 400 ft-lb of work.

Since power is the rate of doing work, it involves time with work.

$$\text{Work} = W \times L$$

where

W = weight in pounds

L = distance in feet, through which W is raised or moved

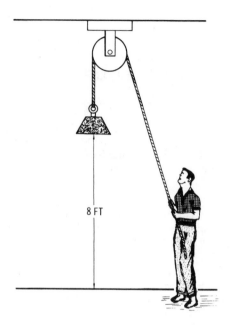

Figure 13-1 Example of work.

8 FT

Horsepower (hp) is a unit of power and, numerically, 1 hp = 33,000 ft-lb per minute or 550 ft-lb per second.

$$hp = \frac{L \times W}{33,000 \times t \text{ (in minutes)}}$$

$$hp = \frac{L \times W}{550 \times t \text{ (in seconds)}}$$

$$P = \frac{W}{t}$$

where

P = power of doing work in foot-pounds per second
W = work done in foot-pounds
t = time of doing work in seconds or minutes, as indicated, or

$$\text{Mechanical Power} = \frac{\text{Mechanical Work Done}}{\text{Time of Doing Work}}$$

Problem

A 1000-lb load was carried by an elevator up 100 ft in 30 seconds. How many horsepower were required to accomplish this? Neglect efficiency.

$$\text{hp} = \frac{L \times W}{33,000 \times t \text{ (in minutes)}} = \frac{L \times W}{550 \times t \text{ (in seconds)}}$$

$$= \frac{100 \times 1000}{33,000 \times 0.5} = 6+ = \frac{100 \times 1000}{550 \times 30} = 6+$$

Table 13-1 gives some very useful units of conversion. Not only may mechanical power be converted into electrical power, but also some others, which will be useful. To convert mechanical power (hp) to electrical power (watts), use the following formula:

Electrical Power (watts) = Mechanical Power (hp) × 746 watts

Table 13-1   Conversion Table of Power

Unit	Watt	Kilowatt	Horsepower	Foot-Pounds per Second	Btu per Second
1 W	1.00	0.001	0.00134	0.737	0.000948
1 Kw	1000.0	1.0	1.34	737.0	0.9480
1 hp	746.0	0.746	1.0	550.0	0.7070
1 ft-lb/sec	1.36	0.00136	0.00182	1.0	0.00129
1 Btu/sec	1055.0	1.055	0.415	778.0	1.0

Efficiency is the measure of output to input. Since one never receives something for nothing, efficiency is always less than 1. All transformations are accompanied by losses.

A list of losses accompanying transformation of electrical power into mechanical power is listed below. Some of these losses have not been covered as yet, since they are involved in alternating current, but they will be listed for reference.

I^2R loss (copper loss)

Friction loss (bearings)

Air resistance

Hysteresis

Eddy currents (will be covered with alternating currents)

Power factor (will be covered with alternating currents)

Reluctance of air gap

Losses are listed here to illustrate why 100 hp is not received from an electrical motor with 74,600 watts input.

A 100-hp DC motor at 240 volts draws 341 amperes. Since $P = EI$, $P = 240 \times 341 = 81,840$ watts. Now, 100 hp equals 74,600 watts, so there is a loss of 7240 watts. The efficiency of the DC motor is output/input = 74,600 watts/81,840 watts = 91.15 percent.

Refer to the elevator problem (p. 000) and convert it into electrical power

Problem

What horsepower of electric motor will it take to lift 1000 lbs a height of 100 ft in 30 seconds? Use an 80 percent efficiency factor. How many kilowatts will be required to supply this horsepower?

$$\text{Mechanical hp} = \frac{L \times W}{550 \times 330} = \frac{100 \times 1000}{16,500} = 6.06 \text{ hp}$$

$$\frac{\text{hp}}{\text{eff.}} = \frac{6.06}{0.8} = 7.6 \text{ hp} \qquad \text{Ans. (1)}$$

$$1 \text{ hp} = 0.746 \text{ kW} \quad \text{so} \quad 7.56 \times 0.746 = 5.64 \text{ kW}$$
$$\text{Ans. (2)}$$

The following are useful formulas to be used with efficiency:

$$\text{Efficiency} = \frac{\text{Output}}{\text{Input}} = \frac{\text{Output}}{\text{Output} + \text{Losses}} = \frac{\text{Input} - \text{Losses}}{\text{Input}}$$

$$\text{Input} = \frac{\text{Output}}{\text{Efficiency}}$$

$$\text{Output} = \text{Input} \times \text{Efficiency}$$

The following formulas are listed for reference:

$$\text{hp} = \frac{\text{Watts}}{746}$$

Watts = hp × 746

$$hp = \frac{kW}{0.746}$$

kW = hp × 0.746

Problem
How many horsepower are 2460 watts?

$$hp = \frac{Watts}{746} = \frac{2460}{746} = 3.3 \text{ hp}$$

Problem
A motor draws 30 kW. How many horsepower is this?

$$hp = \frac{kW}{0.746} = \frac{30}{0.746} = 40.2 \text{ hp}$$

hp = kW × 1.34 = 30 × 1.34 = 40.2 hp

Torque

Applied torque is a measure of a body's tendency to produce rotation. *Resisting torque* is the tendency of a body to resist rotation.

A definition of torque is a twisting or turning force that tends to produce rotation, as of a motor.

Torque is measured by the product of the force and the perpendicular distance from the axis of rotation to the line of action of the force:

$$T = F \times L$$

where

      T = torque turning or twisting efforts in pounds-feet
      F = unbalanced force exerted to produce rotation
      L = lever arm length

Torque is usually expressed in pound-feet. Be careful to observe that this is *not* foot-pounds.

An example of torque is illustrated in Figure 13-2. The torque tending to turn the cylinder in the brick wall would be $T = F \times L$ or $T = 100 \text{ lb} \times 12 \text{ ft} \times 1200$ pound-feet.

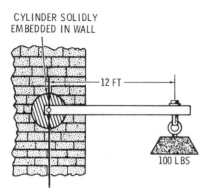

CYLINDER SOLIDLY
EMBEDDED IN WALL

Figure 13-2   Example of torque.

— 12 FT —

100 LBS

Prony Brake

The Prony brake is used for finding the brake horsepower of all types of engines and motors. In running the test, the brake lever is arranged as in Figure 13-3 to bear on a scale S. The pressure F on the scale is regulated by bolt R. The revolutions per minute of the wheel are counted. Here, L is the length of the lever arm. The brake horsepower is calculated as follows:

$$\text{Brake Horsepower} = \frac{2 \times L \times 3.1416 \times F \times \text{rpm}}{33,000}$$

Notice that the product FL in the numerator is the torque of the system.

In electrical units, a watt-hour represents the electrical energy expended if work is done for 1 hour at the rate of 1 watt. By the same token, a kilowatt-hour represents the electrical energy expended if work is done for 1 hour at the rate of 1 kilowatt.

$$1 \text{ joules } = 1 \text{ watt-second}$$
$$3600 \text{ joules } = 1 \text{ watt-hour}$$
$$3,600,000 \text{ joules } = 1 \text{ kilowatt-hour}$$

Electrical energy is purchased and metered on the kilowatt-hour basis. The rates are usually based on a sliding scale. That is, the first so many kilowatt-hours are at a higher rate than the ensuing steps will be. Also, on commercial and industrial power customer rates, the maximum 15-minute demand in continuous kilowatts will set the steps used in calculating the power bill. The reason for this is that if a commercial customer used a continuous demand of

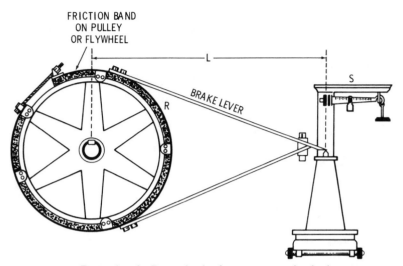

Figure 13-3 Example of a Prony brake for measuring brake horsepower.

10,000 kW, the load on the system would be steady, but if the customer drew 10,000 kW for 15 minutes or an hour, and then dropped to 5000 kW the rest of the time, the power company would have to have a system that could supply the 10,000-kW demand, while the customer used only 5000 kW the majority of the time.

The power company is entitled to a ready-to-serve charge, because we want them to be ready to supply whatever the required demand at any time. The same is true if a company has its own power plant. The investment to serve must be there regardless of whether power is utilized continuously or for only a short time.

Heat may be obtained from electrical energy, both as a loss and as work done. Heat energy is expressed in two ways: calories and Btu (British thermal units).

The *calorie* is the amount of heat required to raise the temperature of 1 gram of water by 1 Celsius degree.

The *Btu* is the amount of heat required to raise the temperature of 1 pound of water by 1 Fahrenheit degree.

Going back to Table 13-1, we find that 1 kW will produce 0.9480 Btu per second.

A current of 1 ampere maintained for 1 second in a 1-ohm resistor produces 0.239 calorie of heat. A current of 2 amperes in the same circuit produces four times the calories (four times the heat).

Multiply 0.239 by the square of the current (in amperes), by the resistance (in ohms), and by the time (in seconds): The result is heat in calories (H):

$$H = 0.239\ I^2Rt$$
$$\text{Calories} = 0.239 \times \text{amperes}^2 \times \text{ohms} \times \text{seconds}$$
$$\text{Btu} = 0.000948 \times \text{amperes}^2 \times \text{ohms} \times \text{seconds}$$

Formulas

$$\text{Work} = W \times L$$

$$\text{hp} = \frac{L \times W}{33,000 \times t\ \text{(in minutes)}}$$

$$\text{Efficiency} = \frac{\text{Output}}{\text{Input}} = \frac{\text{Output}}{\text{Output} + \text{Losses}} = \frac{\text{Input} - \text{Losses}}{\text{Input}}$$

$$\text{Input} = \frac{\text{Output}}{\text{Efficiency}}$$

$$\text{Output} = \text{Input} \times \text{Efficiency}$$

$$\text{hp} = \frac{2 \times L \times 3.1416 \times F \times \text{rpm}}{33,000}$$

$$\text{Calories} = 0.239 \times I^2 \times R \times t$$
$$\text{Btu} = 0.000948 \times I^2 \times R \times t$$

Questions

1. Define energy.
2. Define work.
3. Define power.
4. Define torque.
5. Define efficiency.
6. Give a formula for work.
7. Give two formulas for horsepower.
8. A 2000-lb object is moved 250 ft in 5 minutes. What horsepower will be required for the job?

9. 1 watt = how many hp?

10. 1 hp = how many watts?

11. How many watts equal 1 Btu per second?

12. Name the losses encountered in an electric motor.

13. Theoretical horsepower required for a job is 12. What is the horsepower required if there is 80 percent efficiency involved?

14. Give the formula for torque.

15. Describe a Prony brake.

16. Give the formula used with Prony brake tests.

17. How many joules are in 1 kWh?

18. Give the formula for calories.

19. Give the formula for Btu.

Chapter 14

Instruments and Measurements

Sir William Thompson designed the earliest sensitive galvanometer. This was known as the *Thompson mirror galvanometer*.

Galvanoscopes were mentioned in a previous chapter. The difference between a galvanoscope and a galvanometer is that a galvanoscope detects an electrical current while a galvanometer measures the strength of the current.

There are three classifications of galvanometers:

1. One with a fixed coil and a movable magnetic needle

2. One with a fixed magnet and a movable coil

3. One with a fixed coil and a movable coil and designed as an *electrodynamometer*

Methods are required to keep a weak current from deflecting the moving member as far as a strong current would. Most instruments that you will be using use a spiral phosphor-bronze spring (usually two springs). These springs (see Figure 14-1) are also the electrical connections for the moving coil.

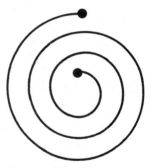

Figure 14-1   Spiral meter control spring.

Not all types of galvanometers will be covered. A lecture table galvanometer is illustrated in Figure 14-2. This uses gravity control. A magnet is suspended horizontally and attached to a pivoted pointer *P* that has a counterweight *W* to keep the bar horizontal. The source of electrical current is connected to terminals *A* and *B*. The pointer *P* is free to swing left or right, depending upon the direction of the current, for a distance proportional to the strength of the current. The scale may be calibrated to read the current directly.

Meters for DC only should never be connected to an AC source. Most of the instruments that you will use in the field are basically DC meters with permanent magnets. When they are designed for use for AC or on DC and AC, there will be a rectifier in the AC circuit

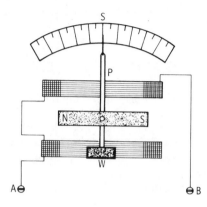

Figure 14-2 Lecture table galvanometer.

to change the current into DC so that this type of movement may be used without damaging the instrument.

A natural question that might arise is Why not connect a DC instrument on AC? Our test instruments are precision equipment and AC demagnetizes permanent magnets. This would, of course, affect the precision of the instrument.

Most DC meters will be damaged when connected to an AC input. Some instruments will give a false indication when used on AC. Most have jeweled movements and should be treated as a fine watch. Have a great deal of respect for your test equipment; the capability of an electrician is easily noted by the way he uses electrical test equipment and other tools.

The D'Arsonval galvanometer, illustrated in Figure 14-3, is the basis for most of the electrical test meters that will be used. A permanent magnet N-S, horseshoe in shape, is mounted with its poles in a vertical position. Mounted vertically between the poles is a soft iron core (A) in a cylindrical form. This concentrates the magnetic flux between the poles. Suspended in the narrow gap between the poles and iron core is a small wire coil (C), which is hung by a fine phosphor-bronze strip (B). This strip controls the coil and also admits current into the coil. The current is removed by a similar strip (D). A spring (E) keeps the tension. This illustration shows the mirror (M) used to deflect a light beam to register on a scale the amount of current. A pointer may be used instead of the mirror to indicate the current values on a scale.

Current admitted into coil C causes it to turn in accordance with Maxwell's rule.

The amount of turning will be in proportion to the current strength in coil C.

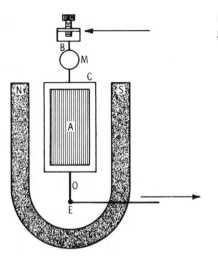

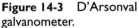

Figure 14-3 D'Arsonval galvanometer.

Voltmeters and Ammeters

Ammeters are in essence galvanometers that read amperes directly on a scale. Voltmeters read volts. At one time, ammeters were basically low resistance and wound with heavy wire, and voltmeters were high resistance and wound with fine wire. (This statement is made in reference to the meter itself.) Now one meter is usually used for combinations of volt-, ampere-, and ohm-reading meters. Voltmeters use resistors in series with the meter to keep the voltage across the meter coil within its limits. Ammeters are connected in shunt (parallel) with low-resistance resistors and read the voltage drop across the shunt resistor. The scales are marked in amperes or volts, whatever the meter is designed to read.

For the most part, many meters are now termed *multimeters*. This indicates that they may be used for volts, amperes, and also sometimes for ohms. This is accomplished by a switch that connects the meter movement internally to the desired connections. *Caution:* Never connect an ammeter across the line, nor connect an ohmmeter to an energized circuit. Always start taking your readings on the highest scale; then move the switch to lower scales as needed.

Edison Pendulum Ammeter

The Edison pendulum ammeter, illustrated in Figure 14-4, consists of a solenoid *A* and a soft iron core *B* pivoted at *C*, with a counterweight *D* to hold the core at the entrance to the solenoid. You may

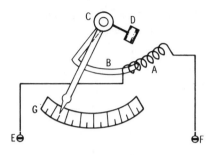

Figure 14-4 Edison pendulum ammeter.

observe that this meter has to be level so the pointer would zero at no current. The iron core will be drawn into the solenoid in proportion to the value of current in *A*.

Battery Gauge

The battery guage is similar to the Edison pendulum ammeter just described, except a phosphor-bronze spring is used instead of the counter weight, and the solenoid has two windings: one of heavy conductor connected to *A* and *B* for the ammeter, and another of fine wire connected to *A* and *C* for the voltammeter (Figure 14-5).

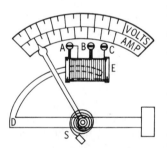

A voltage tester (this is not a voltmeter), the type of which is standard equipment to the electrician, is shown in Figures 14-6 and 14-7.

The Amprobe voltage tester provides an indication of the voltage level. Lightweight, compact units withstand rough usage. Units use neon lamps to

Figure 14-5 Battery gauge.

indicate voltage level, based on the old thermometer style.

As previously stated, most voltmeters and ammeters are of the D'Arsonval movement type. See Figure 14-8. A permanent magnet *M* has two soft iron poles *P* and *P*. Between these poles is a smaller soft iron core *C*, which is supported by brass plates attached to the poles. This core *C* is mounted on a shaft supported by jewel bearings in these brass plates. A coil of fine copper wire *W* is wound around the soft iron core *C* horizontally. There are phosphor-bronze coiled springs, as were shown in Figure 14-1, at each end of the core. These springs control the movement of the core and also serve as flexible leads to supply the electrical current to the moving coil. Maxwell's rule again prevails, and the coil and core tend to turn to embrace the most magnetic flux. A pointer attached to the

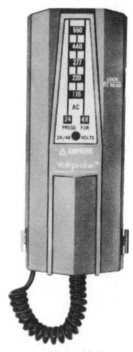

Figure 14-6 Voltage tester. *Courtesy Amprobe Instrument Division of SOS Consolidated, Inc.*

core moves and indicates on a scale the amount of current or voltage. This instrument as illustrated is strictly for DC use. Circuit modification within the meter enclosure is required when used on AC.

The electrodynamometer type of meter utilizes the same basic principle as the D'Arsonval movement, except that the permanent magnet is replaced by fixed coils. This type of movement may be used on either AC or DC without damage since there are no permanent magnets. The DC scale for the same voltage or amperage will be different from the AC scale.

Indicating Wattmeter

Indicating wattmeters are of the electrodynamometer type. They are a combination of a voltmeter and ammeter. The fixed coils are usually the ammeter portion and the moving coil the voltmeter portion. The current coil, being of heavy wire, is the stationary coil. The voltage coil is the movable coil of fine wire and is mounted as in a D'Arsonval movement. At zero watts the voltage coil is at right angles to the current coil. As current and voltage are applied,

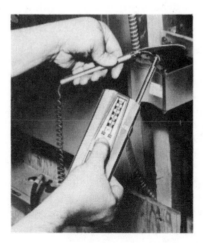

Figure 14-7 Voltage tester in use. *Courtesy Amprobe Instrument Division of SOS Consolidated, Inc.*

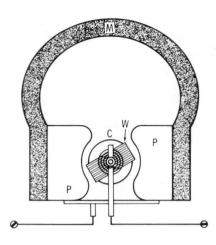

Figure 14-8 Structure of most ammeters and voltmeters.

Figure 14-9
Clamp-on ammeter
and voltmeter. *Courtesy Amprobe Instrument Division of SOS Consolidated, Inc.*

the voltage coil turns (Maxwell's rule) to embrace as many lines of force as possible. The scale reads the product of the current times the amperes, or watts.

A discussion of AC clamp-on ammeters and voltmeters may be a little premature due to the fact that current transformers have not been covered. In order to restrict our discussion of meters to this chapter, they will be shown, however.

Figures 14-6 and 14-7 illustrate voltage testers that use neon lights for voltage indications and may be used on AC or DC voltages. Figure 14-9 shows a clamp-on voltmeter-ammeter. Leads are used to connect the instrument to AC voltage. To test AC amperes, the clamp opens up and encompasses the conductor. It uses the current transformer principle, which will be discussed later with AC. This instrument can't be used on DC.

Figure 14-10 shows a recording meter. Voltmeters, ammeters, and wattmeters come in recording types. These small, compact units provide permanent records of functions being monitored. This frees personnel to do other work while the recorder provides

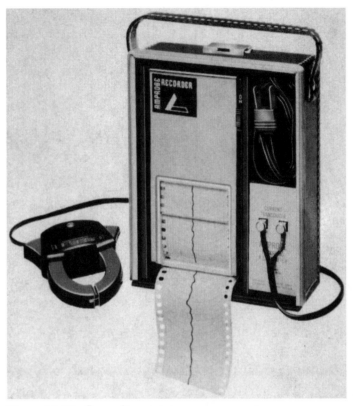

Figure 14-10 **Recording meter.** *Courtesy Amprobe Instrument Division of SOS Consolidated, Inc.*

written records of fluctuation, loads, and on/off operating time, which helps in troubleshooting systems and installations.

Figure 14-11 illustrates a voltmeter connected into a circuit. Figure 14-12 illustrates an ammeter connected into a circuit. Figure 14-13 illustrates a wattmeter connected into a circuit.

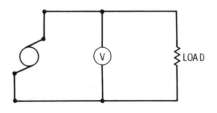

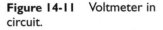

Figure 14-11 Voltmeter in circuit.

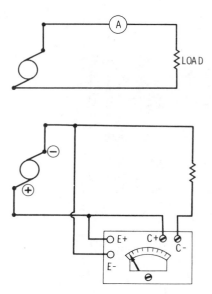

Figure 14-12 Ammeter in circuit.

Figure 14-13 Wattmeter in circuit.

Watts may be measured with a voltmeter and ammeter. This applies to DC circuits only and not to AC circuits. In AC circuits, this would give volt-amperes. This will be explained later, when AC and power factor are studied. For a DC circuit, use a combination of Figures 14-11 and 14-12 and multiply the two readings; thus, 50 amperes × 120 volts = 6000 watts.

Ohmmeters

The portable ohmmeter is an indispensable tool in the electrical trade. It seems that many electricians have never taken time to find out how valuable it is, so they never bother to use one. To the author, they are a necessary crutch. *Caution:* Never connect an ohmmeter to a live circuit, as it will damage the meter and components.

Ohmmeters may be used for continuity checks, resistance checks, and ground checks and save many hours of work. Become familiar with the ohmmeter and it will become as necessary as a pair of pliers.

Periodically all meters should be checked for accuracy against a standard of known accuracy. If a meter is dropped or mishandled, have it checked because an inaccurate meter is of little value.

Thermostats

While thermostats may not strictly be classified as electrical instruments, it is felt that since the thermostat is a device for recognizing differences of temperature, it is appropriate to discuss it. Figure 14-14 illustrates their operation.

Dissimilar metals have different expansion coefficients. Therefore, if brass and steel strips are riveted or brazed together, on a rise in temperature the brass will expand more than the steel, causing the bimetal strip to bend as shown. A drop in temperature will have the reverse effect, so we can use these for furnaces and/or air conditioners.

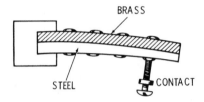

Figure 14-14   Thermostat bimetallic strip.

Thermocouples

When the junctions of dissimilar metals are exposed to a temperature difference, a potential is developed (Figure 14-15). From the electrolysis table of Chapter 9 we find that copper is cathodic as opposed to zinc, so the current will flow from the zinc to the copper.

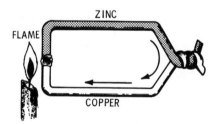

Figure 14-15   Thermocouple.

The thermopile shown in Figure 14-16 is used in the pyrometer. A pyrometer is a meter (microammeter) connected to a thermopile and used to measure high temperatures. The amount of current depends on the temperature. Thus, the microammeter may be calibrated in degrees of temperature and a direct reading made.

Thermocouples are also used in gas furnaces. The thermocouple is heated by the pilot light and the current from the thermocouple

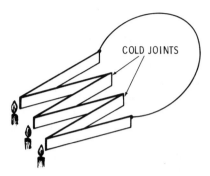

Figure 14-16 Thermopile.

COLD JOINTS

will keep the gas control valve open. If the pilot light goes out, the potential difference will cease and the valve will close, shutting off the gas.

Questions

1. What is the difference between a galvanoscope and a galvanometer?
2. Name three classifications of galvanometers.
3. Sketch a lecture table galvanometer and explain its operation.
4. Why should you not connect a meter that is strictly for DC on AC?
5. Sketch a D'Arsonval galvanometer and tell how it operates.
6. What is the most common movement used in meters?
7. How is the current fed into the core of a D'Arsonval meter?
8. How are "multi" scales developed on a voltmeter?
9. How are "multi" scales developed on an ammeter?
10. Describe an Amprobe voltage tester.
11. In a D'Arsonval meter, what rule that we have learned is used?
12. Describe a wattmeter.
13. How may watts be read in a DC circuit without a wattmeter?
14. Sketch a voltmeter connected into a circuit.
15. Sketch an ammeter connected into a circuit.
16. Sketch a wattmeter connected into a circuit.
17. Describe the operation of a thermostat.
18. Describe a thermocouple and its operation.

Chapter 15

Insulation Testing

There are many insulation resistance testers manufactured by as many different companies. The terms Megged® and Megger® are the registered trade names for insulation testers manufactured by the James G. Biddle Company of Plymouth Meeting, PA.

Earlier, it was stated that there is no perfect insulator. Every insulator to a degree is a conductor of electricity, but most are conductors to a negligible degree. The difference between a conductor and an insulator lies in the electromotive force required to move the free electrons. Dielectrics are insulators.

Dielectric strength is the ability of a dielectric to withstand electrical pressure in volts necessary to puncture the dielectric. It is the maximum potential gradient that a material can withstand without rupture.

The insulation resistance of conductors and all major electrical equipment installed on a new wiring system should be checked, and any faults corrected, before energizing the system, and records should be kept of these tests so that comparisons might be made from time to time to aid in evaluating when preventive maintenance is required. It is also by insulation testing that moisture or dirt may be detected. Insulation testing will expose defects that could be found in no other manner. Wet equipment could be dried, insulation weak spots corrected, or items may be rejected before they are put into service.

Comparison of the original tests with future test records will often reveal weak spots that may be corrected before breakdown.

Megger® insulation testers are DC generators, and the meter is a two-coil type: One is a voltage coil and one is a current coil. The meter is calibrated in megohms and is direct reading as opposed to the homemade tester, which will be described later and requires a formula for computing megohms. If a DC voltage is applied to one or two conductors and the conduit that contains the conductors, a capacitance is set up between the conductors and the conduit. While AC doesn't actually flow through a capacitor, it appears to all practical purposes as though it does. DC will charge the plates and retain the charges. DC will have a charging current that will gradually taper off with a little time and thus won't produce the effect of a current to affect our final reading. Therefore, I feel that a DC tester gives us a more accurate reading.

More will be said later, but first a word of caution. It is suggested that conductors or a conductor and conduit connected to an insulation tester be shorted after a test to bleed off any electrical charge that may have accumulated during the testing. The sizes of the conductors and their length will affect the amount of charge that has been accumulated. Make it a practice to always short out any charges that may have accumulated. It is always better to be safe than sorry. *Don't take this matter lightly.*

The voltage to be used in the test, length of time to run the test, and the acceptable values of insulation resistance will be specified by the design engineer or the inspector and won't necessarily be the values used in this chapter. Whatever the specifications are, follow them.

All manufacturers have specifications for voltages to be used and minimum resistance standards to be met. These are available to you. The values that appear here are representative.

Test Voltages vs. Equipment Ratings

Commonly used DC test voltages for routine maintenance are as follows:

Equipment AC Rating	DC Test Voltage
Up to 100 volts	100 and 250 volts
440–550 volts	500 and 1000 volts
2400 volts	1000–2500 volts and higher
4160 volts and higher	1000–5000 volts and higher

Test voltages used for proof-testing of equipment are considerably higher than those used for routine maintenance. Although there are no published industry standards for DC maximum proof-test voltages to be used with rotating equipment, the schedule given below is customarily used. For specific recommendations consult the manufacturer of the equipment.

Prooftest voltages for rotating equipment:

Factory AC Test = 2 × Nameplate Rating + 1000 volts
DC Prooftest on Insulation = 0.8 × Factory AC Test × 1.6
DC Prooftest After Service = 0.6 × Factory AC Test × 1.6

Example

Motor with 2400-volt AC nameplate rating.

Factory AC Test = 2(2400) + 1000 = 5800 volts AC

Max. DC Test on Installation = 0.8(5800)1.6 = 7424 volts DC

Max. DC Test After Service − 0.6(5800)1.6 = 5568 volts DC

There are three currents that appear in insulation testing:

1. *Capacitance Charging Current.* This is the current that we might say is the capacitor charging current and starts out high, and tapers off with time to zero. See Figure 15-1. No time or current values have been shown, as these are variable. Only a curve that represents them has been shown.

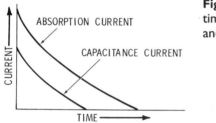

Figure 15-1 Comparison of time and current for capacitance and absorption.

2. *Absorption Current.* Absorption current is due to the polarization of the insulating material. It takes a longer time for absorption current to reach a static condition than it does for the capacitance current to taper off.

It takes a longer time for this charge to bleed off than it does for the capacitance charge. On a long and/or large cable, the large amount of stored energy should be considered and the cable shorted to dissipate the currents after testing. A curve representative of absorption current is also illustrated in Figure 15-1.

3. Conduction or leakage current is a major concern. For practical purposes, this is a steady current due to leakage, both over and through the insulation, caused by moisture, dirt, and normal leakage characteristics of the insulation. Moisture and dirt are generally considered a deterioration of the insulation and cause larger leakage currents.

In discussing moisture, reference is made to the motor, coil, transformer, etc., being wet. Normal moisture in the atmosphere will have no appreciable effect on insulation resistance, unless the temperature is below dew point and actual condensation is present. This won't necessarily hold true where windings have dust on them that is susceptible to drawing moisture from the atmosphere.

In using insulation testers, the test shall be maintained at least 1 minute, or until the reading holds steady for 15 seconds. This ensures that the capacitance charge and absorption charge have reached a static point. This time element will vary with what we are testing; motors and transformers take more than the average conductors, as there are usually more feet of conductor involved.

Figure 15-2 illustrates a Biddle hand-cranked Megger®. Figure 15-3 illustrates a heavy-duty Biddle Megger® in both the hand-cranked and motor-driven types.

Figure 15-2 Biddle hand-cranked Megger® insulation tester. *Courtesy the Biddle Company.*

On 480-volt bus and bus air circuit breakers, use the 1000-volt range on the insulation tester and the minimum acceptable test value shall be 100 megohms.

Cables and Conductors

The insulation of cables and cable and conductors will vary inversely with the ambient temperature, or rather the temperature of the insulation. The insulation resistance will also vary with the type

Figure 15-3   Biddle heavy-duty Megger® insulation tester.
Courtesy the Biddle Company.

of insulation. As the temperature rises, the insulation resistance lowers, and as the temperature lowers, the insulation resistance rises.

Table 15-1 covers several types of insulations. It is one of a number of tables available to you in handbooks. Since temperature affects the insulation resistance, some point in the temperature range must be picked as the zero temperature coefficient. In

Table 15-1   Temperature Coefficient Table

Test Temperature Degrees Fahrenheit	(THW) Thermoplastic	(RHW) Heat Resistant	Butyl Base
50	0.29	0.73	0.70
52	0.40	0.78	0.75
54	0.55	0.83	0.80
56	0.66	0.88	0.86
58	0.82	0.94	0.92
60	1.00	1.00	1.00
62	1.26	1.07	1.06
64	1.55	1.13	1.13
66	2.00	1.20	1.20
68	2.50	1.28	1.27
70	3.10	1.36	1.37
72	4.00	1.45	1.49
74	5.05	1.55	1.58
76	5.95	1.64	1.69
78	7.05	1.75	1.81
80	8.30	1.86	1.94

Table 15-1, this point is 15.56°C or 60°F. With temperatures other than this zero temperature coefficient, the values in the following table must be used to arrive at a proper insulation resistance value.

In testing the conductors (three-phase, and thus three conductors will be used), tie two conductors to the conduit and/or the equipment-grounding conductor as shown in Figure 15-4, and test from the ungrounded conductor and the other two conductors as shown. Repeat to test the remaining two conductors by grounding other conductors and testing the ungrounded conductor, etc.

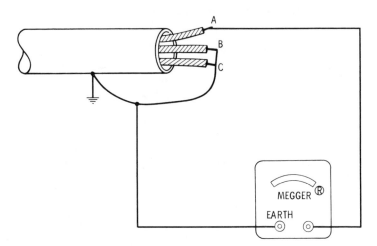

Figure 15-4 Testing feeder conductors.

Take a look at an example using Table 15-1. Assume 600-volt insulation and a 1000-volt insulation tester. The conductor has THW insulation and the ambient temperature is 80°F. We refer to Table 15-1, and find the temperature coefficient to be 8.30. The insulation tester reads the resistance of the insulation at 20 megohms. As stated earlier, a minimum resistance of 100 megohms was required and 20 megohms is well below 100 megohms. But $20 \times 8.30 = 166.00$ when converted back to a base of 60°F, and this is above the 100-megohms minimum requirement. Had this been RHW insulation under the same conditions and reading, it wouldn't have passed the minimum requirements, as the table shows a temperature coefficient of 1.86 at 80°F.

When testing a three-phase AC motor, the author recommends that all three phases should be connected together on the load side of the magnetic starter as shown in Figure 15-5. Table 15-2 is representative of satisfactory tests, when the readings are converted to 60°F, as covered in the formula given below.

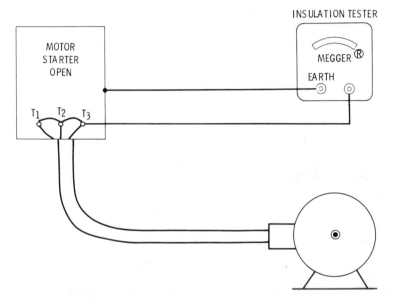

Figure 15-5 Motor insulation test.

Table 15-2   Satisfactory Test Results

Equipment	Insulation Tester Voltage	Minimum Reading in Megohms
460-volt, 3-phase induction motor	1000	20
208-volt, 3-phase induction motor	500	20
120-volt, 3-phase induction motor	500	5

The corrected reading for rotating equipment and transformers shall be in accordance with the following formula:

$$\log R_2 = \log R_1 + \frac{T_1 - T_2}{N}$$

where

$R_1 =$ known resistance at temperature T_1 in degrees Celsius. R_1 is the insulation tester reading in megohms at the time of test and T_1 is the ambient temperature of the winding.

$R_2 =$ unknown resistance of the insulation in megohms at temperature T_2 in degrees Celsius (15.56°C). R_2 is the reading shown in Tables 15-2 and 15-3, and for acceptable reading it should be equal to or greater than the resistance in the tables.

$N = 30$ for class A insulation
$N = 60$ for class B insulation
$N = 23$ for class H insulation

Table 15-3 Acceptable Transformer Readings

Transformer Winding	Insulation Tester Voltage	Minimum Reading in Megohms
480-volt winding	1000	45
277-volt winding	500	45
208-volt winding	500	30

As stated before, the insulation resistance varies inversely with temperature. The changes are not linear but are logarithmic in character.

There are three tests that should be made on transformers. These are illustrated in Figure 15-6: Figure 15-6A, high-voltage winding to ground; Figure 15-6B, low-voltage winding to ground; Figure 15-6C, high-voltage winding to secondary winding. Table 15-3 lists acceptable readings for transformers.

A DC voltage supply such as illustrated in Figure 15-7 may be made up or purchased and used with a double-throw switch, as shown in Figure 15-8, and a high-resistance voltmeter. The formula to be used with this type of insulation tester is

$$R = \frac{r(V - v)}{v(1,000,000)}$$

where

$V =$ voltage at test terminals
$v =$ voltage with insulation in series with the voltmeter
$r =$ resistance of voltmeter in ohms (generally marked on label inside the instrument cover)
$R =$ resistance of insulation in megohms

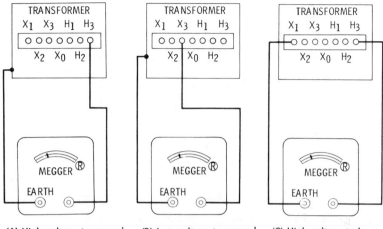

(A) High voltage to ground.   (B) Low voltage to ground.   (C) High voltage to low voltage.

Figure 15-6   Transformer tests.

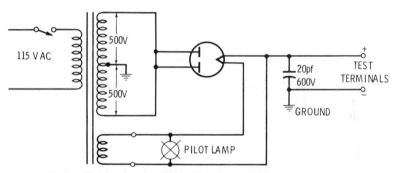

Figure 15-7   Electronic supply for insulation testing.

When drying out wet insulation the resistance will fall rapidly as the temperature is raised during the drying operation. After falling to a minimum for a given temperature, the resistance will gradually rise as the drying progresses and the moisture is expelled from the insulation. See Figure 15-9 for a representative graph, showing a resistance curve and a temperature curve, both plotted against drying time. Take particular note of points *A* and *B*. The resistance is leveling off, indicating that the insulation is dried out and, at this point, the heat is shut off. As the temperature of the winding drops, the insulation resistance will rise rapidly.

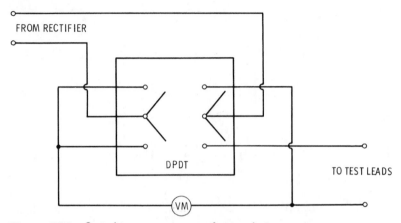

Figure 15-8 Switching arrangement for insulation tester.

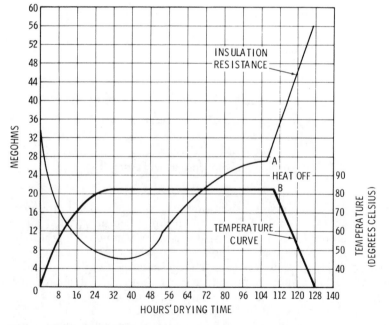

Figure 15-9 Insulation drying curves.

Conductors in Parallel

In Section 310.4 of the *NEC*, conductors in parallel are covered. A number of rules are set out, including the condition that the paralleled conductors shall all have "the same insulation type." At a recent meeting of the Rocky Mountain chapter of the I.A.E.I., the author was asked to explain why the insulations of parallel conductors had to be the same type. This question is often asked and should be answered.

The entire problem in paralleling conductors is to keep the impedance of each conductor as nearly the same as is possible, so the load carried by each conductor will be a balancer. You will recall that impedance is AC resistance, which includes inductive reactance, capacitive reactance, and plain resistance.

Conductor insulation resistance in megohms varies with temperature and types of insulation.

Table 15-1 (p. 000) gives temperature coefficients for three types of insulation. Since there is a great variation in insulation resistance with coefficient temperatures, the leakage of some types of insulation is higher than other types. Leakage adds up to a loss of current, even though it might be a small loss. Thus, if different types of insulation are used on parallel conductors, it will affect the impedances of the paralleled conductors.

A case in point: The author was in charge of checking Megger® insulation tests of large conductors. After pulling in the conductors, an insulation test was to be made. The first test was made in early spring, early in the morning when the temperature was low. Parallel circuits were run across a roof, and in rigid conduit exposed to sunlight. The specs called for RHH-Use to be run. It came time to energize these circuits and insulation tests were run again before energizing. The insulation resistance tested quite low, so the circuits were not energized.

After considerable research and investigation it was found that THW insulation was installed instead of RHH-Use. The temperature was high. Graphs of comparisons of the two types of insulation were made and from these the answer was quite apparent.

Figure 15-10 is a graph that may be used for comparing insulation resistance against temperature for RHH insulation. Figure 15-11 is a graph for insulation resistance coefficient and ties in with Table 15-1 for THW and temperature. (Note that logarithmic graph paper is used.)

From the above information, it may readily be seen why the *NEC* requires the same type of insulation to be used on parallel runs of conductors.

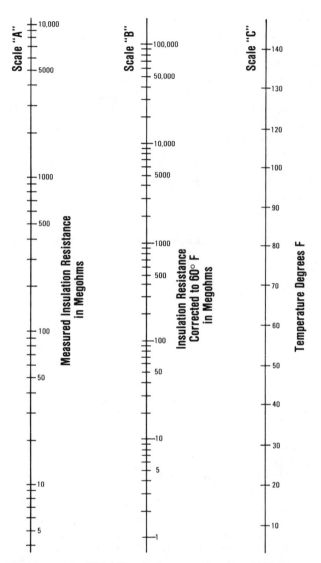

Figure 15-10 RHW insulation corrected to 60°F. Place straightedge on measured Megger® tester value, scale A, and temperature, scale C. Read corrected Megger® value on scale B.

Temperature Degrees F

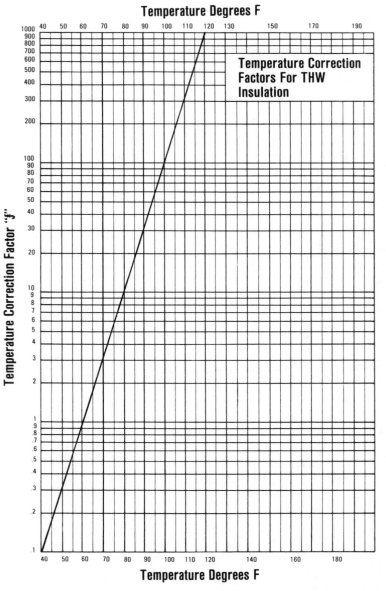

Figure 15-11   Temperature correction factors for THW insulation.

Table 15-4 Temperature Correction Factors*

| Temperature | | Rotating Equip. | | Oil-Filled Trans-formers | Cables | | | | | | | |
°C	°F	Class A	Class B		Code Natural	Code GR-S	Performance Natural	Heat Resist. Natural	Heat Resist. & Perform. GR-S	Ozone Resist. Natural GR-S	Varnished Cambric	Impregnated Paper
0	32	0.21	0.40	0.25	0.25	0.12	0.47	0.42	0.22	0.14	0.10	0.28
5	41	0.31	0.50	0.36	0.40	0.23	0.60	0.56	0.37	0.25	0.20	0.43
10	50	0.45	0.63	0.50	0.61	0.46	0.76	0.73	0.58	0.49	0.43	0.64
15.6	60	0.71	0.81	0.74	1.00	1.00	1.00	1.00	1.00	1.00	1.00	1.00
20	68	1.00	1.00	1.00	1.47	1.83	1.24	1.28	1.53	1.75	1.94	1.43
25	77	1.48	1.25	1.40	2.27	3.67	1.58	1.68	2.48	3.29	4.08	2.17
30	86	2.20	1.58	1.98	3.52	7.32	2.00	2.24	4.03	6.20	8.62	3.20
35	95	3.24	2.00	2.80	5.45	14.60	2.55	2.93	6.53	11.65	18.2	4.77
40	104	4.80	2.50	3.95	8.45	29.20	3.26	3.85	10.70	25.00	38.5	7.15
45	113	7.10	3.15	5.60	13.10	54.00	4.15	5.08	17.10	41.40	81.0	10.70
50	122	10.45	3.98	7.85	20.00	116.00	5.29	6.72	27.85	78.00	170.00	16.00
55	131	15.50	5.00	11.20			6.72	8.83	45.00		345.00	24.00
60	140	22.80	6.30	15.85			8.58	11.62	73.00		775.00	36.00
65	149	34.00	7.90	22.40				15.40	118.00			
70	158	50.00	10.00	31.75				20.30	193.00			
75	167	74.00	12.60	44.70				26.60	313.00			

*Corrected to 20°C for rotating equipment and transformers; 15.6°C for cable.

Effect of Temperature on Insulation Resistance*

The resistance of insulating materials decreases markedly with an increase in temperature. As we've seen, however, tests by the time-resistance and step-voltage methods are relatively independent of temperature effects, giving relative values.

If you want to make reliable comparisons between readings, you should correct the readings to a base temperature, such as 20°C, or take all your readings at approximately the same temperature (usually not difficult to do). We will cover here some general guides to temperature correction.

One rule of thumb is as follows: For every 10°C *increase* in temperature, you halve the resistance, or for every 10°C *decrease,* you double the resistance. For example, a 2-megohm resistance at 20°C reduces to ½ megohm at 40°C.

Each type of insulating material will have a different degree of resistance change with temperature. Factors have been developed, however, to simplify the correction of resistance values. Table 15-4 gives such factors for rotating equipment, transformers, and cable. You *multiply* the reading you get by the factor corresponding to the temperature (which you need to measure).

For example, assume you have a motor with Class A insulation and you get a reading of 2.0 megohms at a temperature (*in the windings*) of 104°F (40°C). From Table 15-4 you read across at 104°F to the next column (for Class A) and obtain the factor 4.80.

Questions

1. How many ohms are in a megohm?
2. Explain capacitive charging current.
3. Explain absorption current.
4. Explain conduction or leakage current.
5. Explain insulation resistance as temperature increases.
6. Explain insulation resistance as temperature drops.
7. Illustrate how to test resistance of conductors.
8. Illustrate how to test resistance of motors.
9. Illustrate how to test resistance of transformers.

*Courtesy the Biddle Instrument Company.

Chapter 16

Electromagnetic Induction

In 1831 Michael Faraday discovered that if a conductor moved across a magnetic flux, so as to cut the lines of force, an emf would be induced in the conductor. This is called electromagnetic induction.

Basic Principles

The conductor may be a straight wire, a coil of wire, or a solid block of metal. It makes no difference whether the conductor moves and cuts the lines of force or whether the lines of force are moving and cut a stationary conductor. Electromagnetic induction ensues in either case. A conductor moving parallel with the lines of force won't cause electromagnetic induction. The conductor must cut or be cut by magnetic lines of force. When the lines of force are cut perpendicularly, more emf will result than if the lines of force are cut at an angle.

The following description of what happens when an electromagnetic induction results is the most meaningful. See Figure 16-1.

In Figure 16-1A the conductor D is moving downward and is about to start to cut a line of force. As the conductor reaches the position in Figure 16-1B, the magnetic line of force is bending. Then, finally, as the conductor reaches point F in Figure 16-1C, the magnetic line of force cuts the conductor. It will be remembered

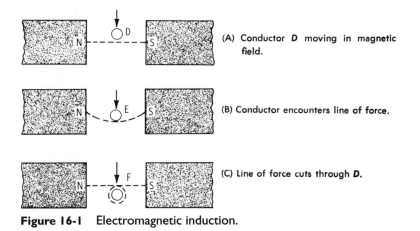

(A) Conductor **D** moving in magnetic field.

(B) Conductor encounters line of force.

(C) Line of force cuts through **D**.

Figure 16-1   Electromagnetic induction.

that when a current from a battery is passed through a conductor, there results a magnetic field around the conductor. If, now, by mechanical means it is possible to pass a magnetic line of force through a conductor, the result will be an emf induced in the conductor.

If, as shown in Figure 16-2, a permanent magnet N-S is rapidly inserted into the coil C, an emf will be induced in the coil and register on the millivoltmeter V. Now, if the magnet N-S is pulled out of the coil C, an emf will be induced again into the coil C, but in the opposite direction.

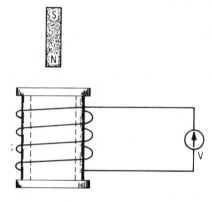

Figure 16-2 Electromagnetic induction with a coil.

The magnitude of the emf depends on the rate of cutting of the magnetic lines of force. One line of force cut by one conductor in 1 second will generate 1 absolute unit of electromotive force. One volt is equal to 10^8 absolute units. From this it can be seen that a conductor must cut 100,000,000 lines of force per second to produce 1 volt.

In Figure 16-2, when the magnet N-S was inserted into coil C, an emf was induced in coil C and current flowed through the voltmeter. This current, in turn, set up a magnetic field in the coil, which set up a magnetic field around coil C. This opposed the pushing of the magnet into coil C. A similar condition occurs when the magnet is pulled out of coil C. A field is set up that opposes the removal of the magnet.

Lenz's Law

Lenz, in 1834, summed up these effects, which are known as Lenz's law: In all cases of electromagnetic induction the direction of the induced emf is such that the magnetic field set up by the resultant current tends to stop the motion producing the emf.

Figure 16-3 will be used along with Figures 16-4 and 16-5 to illustrate self-induction. Self-induction was just described, but not identified as such. If a current is passed through 1000 ft of No. 20 conductor stretched in a straight line in space, and voltage is applied, the current will rise instantaneously in value, as determined by the resistance and applied voltage. Now, if the same conductor is wound into a coil, as in Figure 16-3, and if this coil has a resistance of 10 ohms and the voltage applied is 10 volts, the ultimate current would be 1 ampere: $I = E/R$.

Figure 16-3 Self-induction.

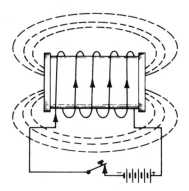

The current won't rise in the coil instantaneously. This is due to the fact that when a current starts flowing in turn A (Figure 16-4), the magnetic lines of force cut turn B and set up a counter-emf opposing the emf applied to the coil. A similar action ensues as all turns are cut in a similar manner. The charging current thus climbs slowly. The switch in Figure 16-3 is closed at point 0 in Figure 16-4. The charging current climbs not instantly, but relatively slowly, as shown by curve

Figure 16-4 Charging current.

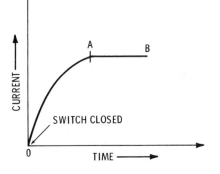

0*A* in Figure 16-4. When the maximum current is reached at point *A*, the current will level off as shown by *AB* in Figure 16-4.

Now if the switch in Figure 16-3 is opened, the turns of the coil have magnetic lines of force around them and this flux will collapse, cutting the turns of the coil, and induce an emf that tends to keep the current going. Thus, there is generated or induced in the circuit an emf opposing any change of the current therein. The generation of this force is called *self-induction,* and it occurs in all circuits that have an inherent property known as *inductance.* Inductance may therefore be defined as an inherent property of a coiled conductor, by virtue of which it opposes any change of current therein. The result of inductance is self-induction. Again, Lenz's law applies. See Figure 16-5.

Figure 16-5 Decay of current.

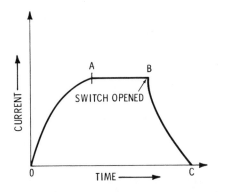

Consider two wires as in Figure 16-6, with current in *A*, and with the resultant magnetic field as shown. Assuming a paralleled conductor *B* moved toward *A*, *B* will indent the magnetic loop about *A*, as shown, causing the flux to circulate around *B* in the opposite direction. This is equivalent to a current flowing in *B* in an opposite direction to that in *A*. The same effect would result if a current were suddenly started in *A*. An emf would be induced in *B* in the opposite direction to that in *A*.

Now consider Figure 16-7; a current is flowing downward in *A*, with a magnetic flux in the same direction as in *A* in Figure 16-6. Parallel to it is conductor *B*, which is now assumed to be moving away from *A*. The magnetic flux is now bent outward and the tendency is for the flux to circulate around *B* in the same direction as in *A*. The same effect would be produced in *B* if the current in *A* suddenly ceased, instead of moving the conductors apart.

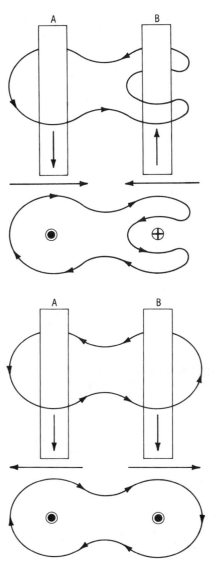

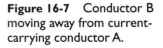

Figure 16-6   Conductor B moving toward current-carrying conductor A.

Figure 16-7   Conductor B moving away from current-carrying conductor A.

Referring to Figure 16-4, the current through the coil has steadied (*AB*). Now the switch in Figure 16-3 is opened, the flux produced by the current in the coil collapses, and a counter-emf is set up attempting to keep the current flowing in the direction that it was when the switch was closed. The current in the coil decays slowly (see Figure 16-5). This will cause an arc across the switch in Figure 16-3.

Fleming's Rule

The emf resulting from the cutting of lines of force may be best understood by *Fleming's rule*. Since current flows from negative to positive, this is a left-hand rule (Figure 16-8). It must be remembered that the thumb represents the direction of motion of the conductor and not that of the flux motion.

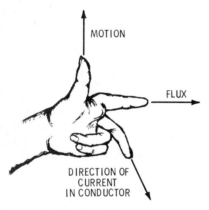

MOTION

FLUX

DIRECTION OF
CURRENT
IN CONDUCTOR

Figure 16-8   Fleming's rule for generation of emf in conductors.

Mutual Induction

So far, this discussion has centered on self-induction. There is also mutual induction. The principles are the same as with self-induction, but we will be dealing with induction caused by one coil inducing an emf into another coil or object.

A simple illustration is shown in Figure 16-9. The circuit is closed by the switch *S*, and the current flows through the conductor

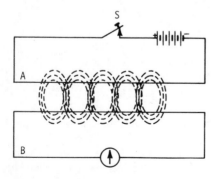

S

A

B

Figure 16-9   Mutual induction.

A setting up magnetic lines of force that cut across *B*, inducing an emf in the opposite direction to the emf acting in *A*. Opening the switch *S* causes the magnetic flux to collapse across *B* and induces a reverse emf.

When electromagnetic induction takes place from circuit *A* (Figure 16-9) to an adjoining circuit *B*, electrically insulated from each other, it is called *mutual induction*.

You will recall that one line of force cutting one conductor in 1 second induces 1 absolute unit of emf. Using coils instead of single conductors, as in Figure 16-10, increases the induced emf. Since air has more magnetic reluctance than iron does, more induced emf will result if the air core in Figure 16-10 is replaced with an iron core as in Figure 16-11.

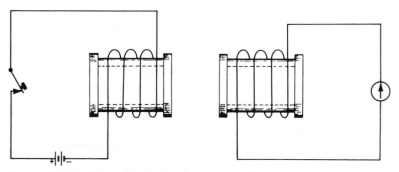

Figure 16-10   Mutual induction between coils.

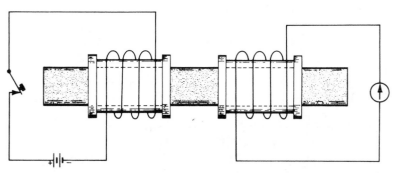

Figure 16-11   An iron core increases the induced emf.

Electromagnetic induction plays a very important role in our lives. A few of the items that wouldn't be possible without induction are

Transformers	Generators
Voltage regulators	Choke reactors
Ignition coils	Radio coils and transformers
Magnetos	Induction motors
Compensators	Watt-hour meters
TV coils and transformers	Alternators

Eddy Currents

Eddy currents are currents due to induced emf and account for some of the losses in electrical equipment. They produce heat. Eddy currents are sometimes harmful, and sometimes they are put to use.

Consider a solid iron rotor of a generator or motor (see Figure 16-12). The solid iron rotor cuts magnetic lines of force as it rotates the same as the coil *C-E-F-D* does. The coil has an emf induced in it in the direction shown. Also, the iron core has an emf induced in it in the same direction as the coil *H-I-J-K*. This will be a low-resistance circuit and the current that results does no work, merely heating the iron core. If the resistance of path *H-I-J-K* were $\frac{1}{1000}$ ohm and the induced emf in the core were 5 volts, we would have 5000 amperes flowing in the core. The heating effect of current is proportional to the square of the current $(P = I^2R)$. The currents in the iron core are commonly called *eddy currents*, but are also known as *Foucault currents*.

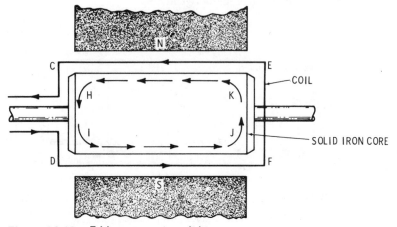

Figure 16-12 Eddy current in solid iron core.

If the core were broken up into sections and these sections were insulated from each other as in Figure 16-13, the induced emf would be broken up into four smaller emfs, cutting the current and heating effects down.

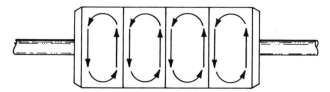

Figure 16-13   Eddy currents in sectionalized rotor.

The way eddy currents are minimized is to laminate the core, that is, subdivide the core into laminations approximately ¼₄ inch thick, with one side varnished for insulating effect. These laminations are assembled on the shaft to make up the iron core. The eddy currents are extremely small compared to those in the solid iron core; thus the losses are cut down.

Eddy currents are influenced by (1) the square of the speed, (2) the square of the flux density, and (3) the square of the thickness of the laminations.

Questions

1. Explain two methods of causing electromagnetic induction.
2. State Lenz's law and explain it.
3. Define an absolute unit of electromotive force.
4. How many absolute units of electromotive force does it take to make one volt?
5. Explain Fleming's rule.
6. What is self-induction?
7. What is inductance?
8. What is mutual induction?
9. How does mutual induction differ from self-induction?
10. Name 10 items in which induction plays an important part.
11. What are eddy currents?
12. What three items influence eddy currents?
13. How are eddy currents minimized?
14. Are eddy currents useful?

Chapter 17

Alternating Currents

Up to this point, alternating current (AC) was mentioned only briefly while emphasis was placed on direct current (DC).

Advantages of AC

AC is used much more extensively than DC. There are reasons for this, and a few of the reasons will be taken up now. However, both AC and DC have their uses and advantages.

One of the principal reasons for the widespread use of AC is the effect that voltages have on the cost of transmission and distribution of electrical energy. As an easy example, let us take 5000 kW that has to be transmitted a distance of x with a 3 percent loss. At 1000 volts, to transmit this 5000 kW a distance of x would require 10,000 lb of copper to maintain the 3 percent loss. Now, if the voltage were raised to 10,000 volts by using a step-up transformer, the same 5000 kW of electrical energy could be transmitted the same distance x with a 3 percent loss and only 100 lbs of copper would be required.

From this it may be stated that for the same quantity of power and the same percentage of loss for transmitting electrical energy the same distance, the weight of copper required varies inversely with the square of the voltage.

High-voltage DC is used occasionally for transmission. There are problems involved that no doubt will be overcome. To change DC from one voltage to another, it is necessary to use a motor-generator set, or convert to AC and transform the voltage required and then convert back to DC by means of rectifiers.

With AC, transformers may be used to transform the voltage, either up or down, and the efficiencies of transformers are probably as high as the efficiency of any other electrical equipment.

With AC, the electrons flow in one direction for one-half a cycle or one alternation, and then flow in the other direction for the next alternation. See Figure 17-1.

Generation of AC

Alternating-current generators, which are commonly known as *alternators,* are used to generate AC voltage. In the next chapter you will find that DC generators use the same principle with modifications.

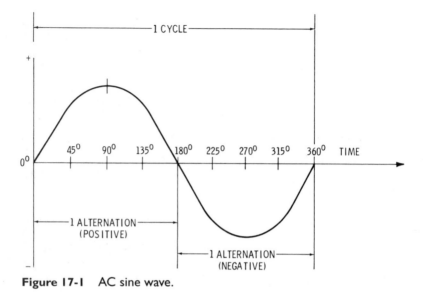

Figure 17-1 AC sine wave.

As a rule, an alternator has the AC winding stationary and this is termed the *stator*. The rotating portion, or *rotor*, is the DC part of the alternator and receives its voltage and current from an exciter that is generally mounted on the alternator and the armature of which is driven by the rotor shaft of the alternator.

Small alternators often use permanent magnets for the rotor, since alloys have been developed that have high magnetic properties. Some small alternators have one winding on the rotor for AC and DC. The AC is taken off by means of slip rings and brushes and the DC by means of a commutator and brushes.

Large alternators usually will generate from 13,800 to 25,000 volts. The reason that the stator is the AC portion is that it is much easier to insulate for high voltages in the stationary portion of the alternator. The DC part is low voltage, so insulation is no problem in the rotor.

You will recall that with induction the magnetic field may move, causing the magnetic lines of force to cut the conductors; or the conductors may move, cutting the magnetic lines of force.

In the explanation of AC generation, the DC portion will be the stationary portion, and although there is usually a DC winding, this won't be shown to simplify the drawings. The AC portion will be the rotating part of the alternator. One turn only will be shown. Many turns are required to generate much of a voltage, but would complicate the explanation. Figure 17-2 will be used with the discussion.

In Figure 17-2A, a one-turn coil is in position No. 1, which is 0°; it is not cutting the magnetic lines of force. The voltmeter (M) is connected to the coil by means of slip rings and brushes. Notice that the voltmeter is registering zero volts. This is because no magnetic lines of force are being cut. In Figure 17-2B, the coil has moved 90° and is cutting the most magnetic lines of force, so the voltmeter has moved to the right, indicating maximum voltage. In Figure 17-2C, the coil has turned 180° from Figure 17-2A and the

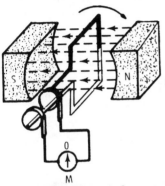

(A) Position No. 1: 0°.

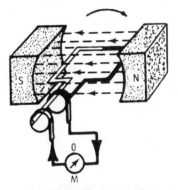

(B) Position No. 2: 90°.

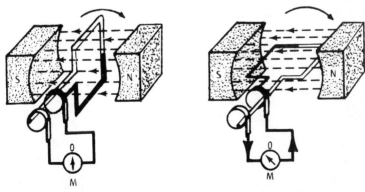

(C) Position No. 3: 180°.

(D) Position No. 4: 270°.

Figure 17-2   Action of alternator rotating coil.

meter again registers zero, as the coil is not cutting any magnetic lines of force. Then, in Figure 17-2D, the coil has turned 270° from Figure 17-2A and again is cutting maximum magnetic lines of force, but in the reverse direction from what it was in Figure 17-2B. Therefore, the voltmeter shows maximum voltage but in the reverse direction from what it showed in Figure 17-2B. The next quarter-turn brings the coil into the position we started with in Figure 17-2A, that is, it has turned 360°.

Now, if you refer to Figure 17-3, you will see the coil rotation in relation to an AC sine wave, or one cycle. The degrees of rotation are noted, and the development of the sine wave can be seen. The induced voltage goes from zero to maximum in one direction, then to zero, and then to maximum in a reverse direction, and back to zero at the completion of the 360° rotation.

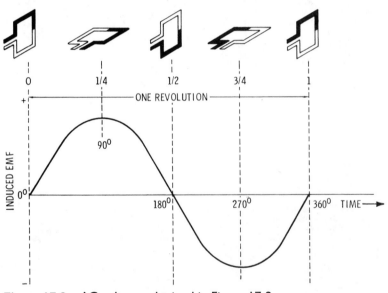

Figure 17-3 AC voltages obtained in Figure 17-2.

The first portion of this sine wave, from 0° to 180°, is one alternation, and the second portion of the sine wave, from 180° to 360°, is another alternation. Thus, two alternations complete one cycle.

The frequency of AC is the number of cycles per second (hertz), and the number of alternations per second is twice the cycles per second.

The frequency of 60 hertz (or cycles per second) is most common in this country, but 50 hertz is used in England and Canada, and 25 hertz is used on some railway equipment.

AC Sine Wave

The AC sine wave demands considerable analysis. First, note that at two times per cycle, the voltage and current pass through zero. Next, when you are using a voltmeter or ammeter, just which values of the wave are being read?

First, looking at Figure 17-4, we find the maximum voltage or current at the points *A* and *B*. An analysis of the sine wave brings one to the conclusion that the maximum just can't be what we would read, because the voltage and/or current continually vary between zero and maximum. Therefore, there just has to be some value that instruments can be manufactured to read.

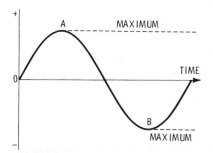

Figure 17-4 Maximum or peak values of a sine wave.

The values that we read on ordinary voltmeters and ammeters are the *effective* or *root-mean-square* (rms) values.

What does *rms* mean? If a near-infinite number of lines are drawn from the zero line to the sine wave, as illustrated in Figure 17-5, and

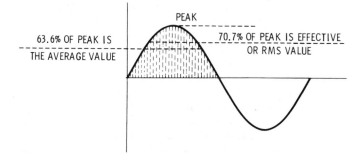

Figure 17-5 Peak or maximum, effective or rms, and average values of sine wave.

each value of each line is squared, these squares added together, and this sum divided by the number of lines and then the square root of this answer taken, we would find this to come out to 70.7 percent of the maximum value. This 70.7 percent of the maximum is what ordinary voltmeters and ammeters will indicate. Take 1.414 amperes or volts as maximum value and multiply this by 70.7 percent and our answer will be 1. So for every volt read on an AC voltmeter, the system is being subjected to a maximum of 1.414 times that voltage.

To determine the power in watts from 1.414 times maximum voltage and current we don't multiply 1.414 by 1.414 to get P, but use 70.7 percent of these figures, that is, the readings on the meters, $P = E_{read} \times I_{read}$. There are meter-type instruments that will measure the maximum or peak voltages, but they are not in common use by electricians.

Another term used is *average* voltage or current. The average voltage or average current is arrived at by taking the average of the values of the lines shown in Figure 17-5, without involving the squares of the values or the square root of the average of the squares. This value (average) equals 63.6 percent of the maximum value.

The maximum value of voltage is of interest to us since if our voltmeter reads 1000 volts, the system is being subjected to 1414 volts twice per cycle.

The heating effect or power output of AC is something that concerns us. Consider a DC circuit with 1 ampere flowing through a resistor R, causing R to heat to 1000°F. Now suppose there is 1 ampere maximum flowing in an AC circuit through the same resistor. The 1-ampere maximum produces only 707°F in resistor R as opposed to 1000°F in the DC circuit. This indicates that only 70.7 percent of the maximum current is effective in an AC circuit.

The theory of the formation of a sine wave will now be explained. In Figure 17-6, conductor A is not cutting magnetic lines of force, which corresponds to position A on the sine wave. The coil rotates 22½° counterclockwise to position B. The time scale for 22½° to the right is F, and where F and B intersect on the sine wave voltage is induced at the 22½° position. Next, conductor C moves to 45°, and on the time line this is point G. Next, the conductor moves to the 67½° point at D. Projecting horizontally from D to right, and vertically from the 67½° mark on the time line, is point D on the sine wave. Next, the conductor moves to the 90° point E. Projected horizontally to the right, and vertically from the 90° mark on the time line, is point E on the sine wave, or the maximum point.

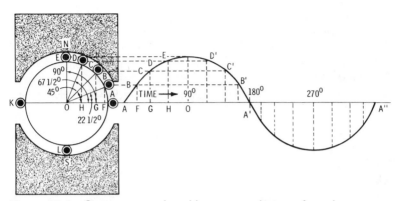

Figure 17-6   Sine wave produced by one revolution of conductor on armature in a bipolar field.

Likewise, by continuing the rotation of the conductor counterclockwise, points D', C', B', A', etc., can be located, until a complete revolution of 360° has been made.

Bear in mind that electrical degrees are being discussed. This doesn't necessarily mean one complete revolution; and if it did mean one revolution, it would be a bipolar alternator that was being considered. In a four-pole alternator, two N and two S poles, one-half of a revolution would be 360 electrical degrees. For alternators with six poles, one-third revolution would be 360 electrical degrees. In essence what is being said is that one cycle is 360 electrical degrees, two cycles are 2 × 360 electrical degrees, etc.

Paralleling Alternators

On our modern electrical systems, many alternators are connected in parallel to the same power grid or transmission system. This allows the best use of the electrical energy outputs of the alternators, as peak loads in various areas will vary, plus an alternator(s) may be shut down during low peak times.

In order to parallel alternators, not only must the voltage outputs be the same, but the phase relationships must be the same (which we will cover more thoroughly later) and the rotating members of the alternators must be turning identically, pole for pole. Note that it was not said that they had to be turning at the same rpm (revolutions per minute), but pole for pole. In simpler terms, when conductors are cutting the magnetic fields of each alternator, they must be at the same angle of electrical degree. Otherwise they won't parallel; or, to use the common phrase, they won't synchronize.

The entire system must be controlled to give exactly 60 hertz, as we have come to rely on our power system to give us the correct time on our electrical clocks.

Formulas

Frequency is in hertz. Formulas commonly used to find frequency, speed, or number of poles are

$$F = \frac{NP}{120}, \quad N = \frac{120F}{P}, \quad \text{and} \quad P = \frac{120F}{N}$$

where

F = frequency in hertz
N = speed in rpm
P = number of poles

To illustrate: Take a two-pole alternator producing 60 hertz. What speed will it have to be driven?

$$N = \frac{120 \times 6}{2} = \frac{7200}{2} = 3600 \text{ rpm}$$

Another example: A four-pole alternator, producing 60 hertz, will have to be driven at how many rpm?

$$N = \frac{120 \times 60}{4} = \frac{7200}{4} = 1800 \text{ rpm}$$

The next few chapters will digress from AC, but will come back to it a little later. It was covered here, as it makes the explanation of DC generation simpler.

In closing this chapter, remember that some form of prime mover must be used to turn the alternator, and thus convert some form of energy into electrical energy.

Questions

1. Explain some advantages of AC over DC.
2. Draw a sine wave and explain its formation.
3. What is an alternator?
4. Is the AC portion of an alternator the stator or rotor?

5. Sketch five positions of a single-coil alternator and compare the positions to the points on a sine wave.

6. What is a hertz?

7. What is an alternation?

8. What is frequency?

9. Explain the difference between degrees of a circle and electrical degrees.

10. What is maximum AC voltage?

11. What is effective AC voltage?

12. What is rms voltage?

13. What is average voltage?

14. What percent of maximum voltage is rms voltage?

15. What voltage does a voltmeter read?

16. A 60-hertz alternator has eight poles. What is its speed?

Chapter 18

DC Generators

Practically all generators produce alternating electromotive force in their windings. This is true whether it is an alternator, as covered in Chapter 17, or a DC generator. This is inevitable, in view of the principle of electromagnetic induction that is involved.

Generators vs. Alternators

As opposed to alternators, the stator supplies the field of the magnetic lines of force. The rotor, or *armature,* as it is called in a DC generator, is the part in which the emf is generated.

In alternators the coils are terminated with slip rings and brushes. In DC generators a means must be provided to collect the induced emf in a manner such that the emf will be taken from the generator in one direction only, instead of an alternating direction.

This is accomplished by means of a commutator, as illustrated in Figures 18-1 and 18-2. Figure 18-1 is a cross-sectional view showing the various parts of a typical commutator. Figure 18-2 shows the coil end of a commutator. This illustration shows the slots into which the conductor ends of the coils are placed and soldered. Each commutator segment is insulated from the next segment by means of mica. Mica has proved to be a very good insulator for commutators.

Figure 18-3 is quite similar to the figures used in Chapter 17 to illustrate alternators. The difference here is that a commutator is used instead of slip rings. In this case, the commutator may be likened to one slip ring that has been split lengthwise and the two halves insulated from each other.

Generation of EMF

In Figure 18-3A and 18-3C, no emf is being induced, as no lines of force are being cut. In Figure 18-3B, the direction of the emf is reversed in the coil from that in Figure 18-3D. Notice, however, that due to the commutator, the emf in Figure 18-3B is going to the voltmeter in the same direction as it is in Figure 18-3D, regardless of the reversal of the emf directions in the coils.

The waveform that we get from a single DC generator is shown in Figure 18-4. Notice that the emf reaches zero twice in one revolution. This waveform has been identified in relation to Figure 18-3. It is an extremely fluctuating DC output, which is not very practical.

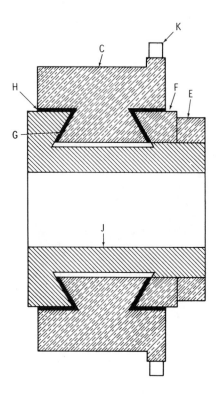

C= COMMUTATOR SEGMENTS (COPPER
G & H= MICA INSULATION
F= BEVELED METAL RING
E= NUT TO HOLD F IN ON BARREL
J= METAL BARREL
K= LUG OR NECK FOR CONDUCTOR
   CONNECTION

Figure 18-1 Sectional view of commutator showing shape and arrangements of segments and insulating material.

Armatures

Figure 18-5A illustrates a double-coil armature. Note that both coils are connected to the same two commutator sections. This double-coil armature will give a waveform as shown by the solid line in Figure 18-5B. The dashed lines are cut off by the commutation.

When a multicoil armature is used, such as would require a commutator similar to that illustrated in Figure 18-2, the output would be practically a straight line output as shown in Figure 18-6. The dashed-line portions of the wave that are shown extending down to

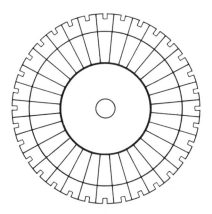

Figure 18-2 Coil end view of commutator showing slots for coil connections.

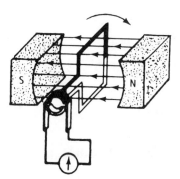

(A) Position No. 1.

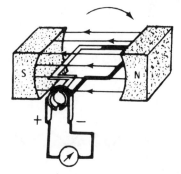

(B) Position No. 2.

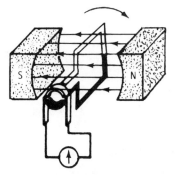

(C) Position No. 3.

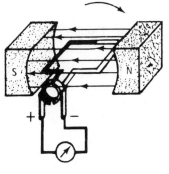

(D) Position No. 4.

Figure 18-3 DC generator coil.

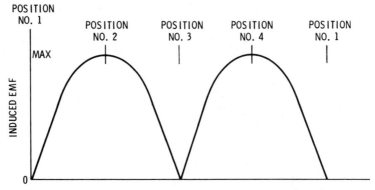

Figure 18-4   Emf from a single-loop coil showing voltage fluctuation.

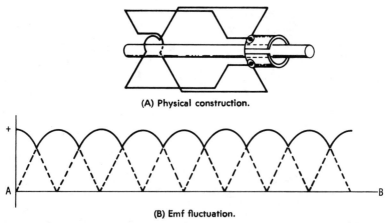

(A) Physical construction.

(B) Emf fluctuation.

Figure 18-5   Double-coil armature.

Figure 18-6   Emf from a multicoil armature.

line *AB* are cut off, so a practically nonpulsating DC emf is produced by the generator.

The DC field poles need not be laminated, but since the armature in reality has AC in it, the armature must be laminated to keep the eddy currents to a minimum.

The steel in the armature laminations must be of the softest and most permeable grade, to keep the hysteresis losses to a minimum.

In all generators, both AC and DC, copper losses, which were covered in preceding chapters, are ever present. The I^2R loss, which is the copper loss in the armature due to current and resistance, produces nothing but a heat loss in the armature.

There is also the *IR* drop or voltage drop in the armature. This is to say that the armature has higher emf induced in the winding than is received at the commutator brushes whenever current is being drawn. At no load will the two voltages be the same. As load is applied, the *IR* drop will vary as the ampere load increases or decreases.

Brushes

It will be noticed when looking at a DC generator that the brushes are practically never in a zero position, such as shown in Figure 18-7A, but are moved slightly in the direction of the armature rotation as shown in Figure 18-7B. This is because, when in operation, the magnetic flux is shifted slightly in the direction of rotation by armature reaction, so what would be a zero position for the brushes is also shifted. The shifting of the brushes reduces sparking at the brushes and the burning of the commutator segments as well as the brushes. Commutating poles are added to the field to counteract armature reaction. These are series field poles, located between the shunt field poles, as shown in Figure 18-8.

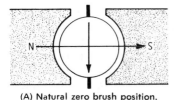

(A) Natural zero brush position.

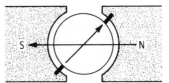

(B) Brush shift position under loads.

Figure 18-7 Brush positions.

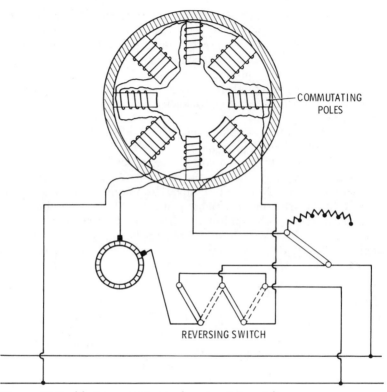

COMMUTATING
POLES

REVERSING SWITCH

Figure 18-8 Motor-reversing and commutator poles.

Types of Generators

There are series, shunt, and compound generators. The design for which the generator is to be used dictates which of these three types would be the most practical for the application. Regardless of the design, DC generators are self-exciting. This is to say that the DC field receives its current from the armature. One may ask: With soft iron or steel being used, where does the magnetism at start-up come from? This magnetism is residual magnetism. All iron and steel, even soft iron and steel, retains a slight amount of magnetism after the magnetizing current is removed. So this residual magnetism in the field poles starts inducing the voltage in the armature windings, which in turn adds more magnetism to the field, and so on until the field strength reaches operating values.

There is a slight chance that all magnetism could be lost. If this ever happens, an external source of DC will be needed to start the generation. DC is connected to the field to give a flash buildup of magnetism. The chances of this happening are remote, but mention has been made of this possible condition, so the electrician may more readily diagnose the trouble.

Figure 18-9 illustrates a series generator. This was used a great deal with the old series street lighting. The field strength increases with the load. Series motors are still used to some extent. The universal motors used in electric saws, drills, mixers, vacuum cleaners, etc., are series motors and may be used on AC or DC current.

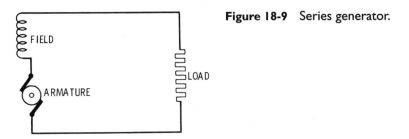

Figure 18-9 Series generator.

Figure 18-10 illustrates a shunt generator. This is well adapted for maintaining a fairly constant voltage under varying loads.

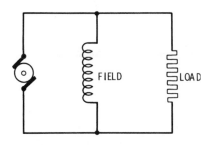

Figure 18-10 Shunt generator.

Figure 18-11 illustrates a compound-wound generator, which is a combination of a series generator and a shunt generator. The series

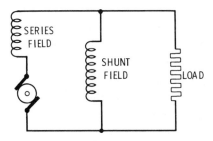

Figure 18-11 Compound generator.

winding aids in maintaining constant voltage with variable current output. As the load goes up, more current goes through the series field winding and raises the voltage to take care of the voltage drop.

Generator Voltage Control

The voltage output of a generator is controlled by the field strength. Figure 18-12 shows a hand-controlled rheostat R in series with the shunt field. (A rheostat is a variable resistor.) As resistance is cut out by moving the arm, more voltage is applied to the shunt field, raising the voltage output of the generator.

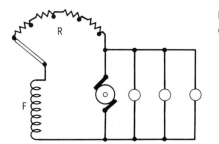

Figure 18-12 Hand voltage control.

Automatic voltage regulators are used on large DC generators. These automatically cut out or cut in resistance into the shunt field circuit as required.

Questions

1. Describe a commutator in your own words.

2. Sketch a multisegment commutator and identify the parts.

3. AC voltage is induced in the winding of a DC generator. True or false?

4. In the zero brush position, are the brushes always at right angles to the field poles? Explain.

5. In which direction are the brushes shifted to reduce sparking?

6. Draw a schematic of a series generator.

7. Draw a schematic of a shunt generator.

8. Draw a schematic of a compound generator.

9. How is voltage regulation obtained in a shunt generator?

Chapter 19

DC Motors

There is no difference in the basic electrical and mechanical designs between DC generators and DC motors. A DC generator will "motor," and a DC motor will "generate." Of course, both devices are generally fine-tuned for their applications, but the basic design is the same.

There were many and varied types of DC motors developed in the early 1800s. From all of these the present-day design evolved.

Armature Movement

Figure 19-1 illustrates how the current-carrying conductors of a DC motor armature tend to move in a magnetic field. Conductor A carries the current away from the observer. From the left-hand rule, the flux in A is counterclockwise. In B the current is flowing toward the observer and, again from the left-hand rule, the flux around B is clockwise. The flux from the field is going from north pole (N) to south pole (S). Thus, A is repelled upward by the field flux and B is repelled downward, so the armature moves in a clockwise direction.

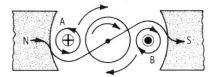

Figure 19-1 Conductors carrying direct current tend to move in a magnetic field.

DC Motors vs. DC Generators

For comparison of a DC motor and a DC generator, Figures 19-2 and 19-3 will be used. The generator in Figure 19-2 is being driven counterclockwise and according to Lenz's law, the reaction of the induced

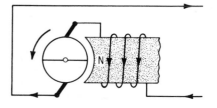

Figure 19-2 Series DC generator: direction of current and rotation.

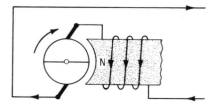

Figure 19-3   Series DC motor: direction of current and rotation.

currents tends to oppose the motion that produces the currents. Thus, if the armature is forcibly driven as shown, the magnetomotive force of the armature, in its reaction on the magnetic field, will oppose the direction of rotation.

Now, if current is applied to the generator, it becomes a motor, as in Figure 19-3. The external driving force having been removed, the armature is free to motor in obedience to the magnetic lines. It will rotate in the opposite direction (clockwise) to which the generator was driven.

The same results occur with shunt- and compound-wound machines. Thus a generator arranged to run as a motor will run opposite to the direction it was driven as a generator.

Referring to generators illustrated in Figure 18-7, the field shifts in motors and the brushes are in a different direction. With a generator the brushes are shifted in the direction of rotation, and with a motor they are shifted in the opposite direction to the direction of rotation. Figure 19-4 illustrates this.

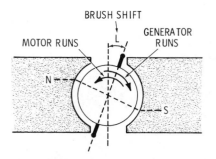

Figure 19-4   Relative directions of a motor and a generator, with armature and field polarities the same.

Regulation of DC Motors

In Figure 19-1, conductor *A* moves up as the motor rotates, but in so doing, *A* also cuts magnetic lines of force; so an emf is induced in conductor *A* (armature conductors), and this induced emf opposes the motion that produces it. So the induced emf is opposite to the emf that is applied to run the motor.

The emf delivered to the motor is called the *impressed voltage*. The opposing emf is called the *counter emf* (cemf), and the difference between the two voltages is called the *effective emf*. Figure 19-5 will be used to illustrate the regulation of a shunt motor.

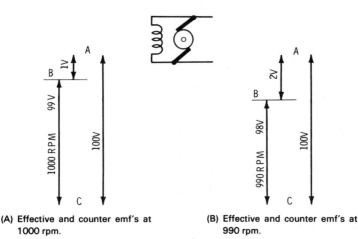

(A) Effective and counter emf's at 1000 rpm.

(B) Effective and counter emf's at 990 rpm.

Figure 19-5 DC shunt-motor regulation.

The resistance of an armature is very low. For this discussion an armature resistance of 0.1 ohm will be used. At standstill the current drawn by the armature will be $I = E/R = 100/0.1 = 1000$ amperes, which, if the current stayed at this value, would burn up the armature winding. When the motor reaches 1000 rpm, there will be a counter emf of 99 volts induced; so the effective emf will be $100 - 99 = 1$ volt. Again $I = E/R = 1/0.1 = 10$ amperes. More load is applied to the motor and this slows the rotation down some, to, say, 990 rpm. Now the counter emf will drop to 98 volts, so $100 - 98 = 2$ volts effective, and $I = E/R = 2/0.1 = 20$ amperes effective to allow the motor to pick up the additional load.

A series motor varies widely in speed with variations in the load. The motor's field strength varies with the load as the armature current will vary with the load; and the armature and field are in series, so they both receive the same current. Also, the combined resistance of the field and armature is higher than the resistance of the armature of a shunt motor.

In Figure 19-6A, assume a speed of 1400 rpm at a given load. Line *AC* is the line voltage of 100 volts, and *CB* is counter emf of 90 volts, so the effective emf will be 10 volts. Assume that the armature and field resistance are 1.0 ohm. Under this condition, $I = E/R = 10/1 = 10$ amperes.

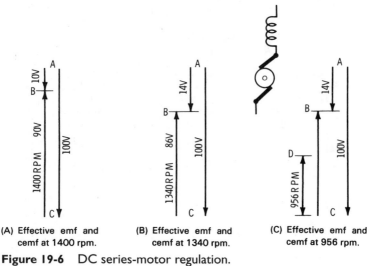

(A) Effective emf and cemf at 1400 rpm.

(B) Effective emf and cemf at 1340 rpm.

(C) Effective emf and cemf at 956 rpm.

Figure 19-6   DC series-motor regulation.

The motor load is increased, so the armature slows down a little. The torque demand of the load has doubled. If the field strength were constant as with a shunt motor, the armature current would have to double. Since an increase in armature current in a series motor also increases the current in the field to the same value, to get double torque the current needs to increase only about 40 percent.

If the field current and the armature current are 10 amperes in the shunt motor at a certain torque, the produced torque is proportional to $10 \times 10 = 100$. If the load doubled, the field would still be 10 amperes, so the current in the armature would have to double to 20 amperes. Then $10 \times 20 = 200$, which would be the measure of the torque required for double the load. Noting the above, with the series motor a 40 percent increase in current was required to get double torque for the doubled load. Remember that the field current also increased by 40 percent. Thus, in Figure 19-6B, with double load the field current would be 14.1 amperes and the armature current would also be 14.1 amperes. Then, $14.1 \times 14.1 = 200$, the required torque for double the load.

In order to clarify things, use Figure 19-6C and for the moment assume that the field strength of a series motor doesn't vary with the armature current, but of course it really does. The 40 percent additional current would require the counter emf in Figure 19-6A to cut from 90 volts to 86 volts, as in Figure 19-6C. This would allow the effective emf to increase from 10 volts to 14 volts, so I = effective E/R or 14/1 = 14 amperes. The speed would fall in direct proportion with the counter emf; thus 90 : 86 = 1400 : x. Then x would equal about 1337 rpm.

The 14 amperes flow through the field windings, however. Assume the field poles are not saturated, so with an increase in flux of 40 percent, say, from 1,000,000 lines to 1,400,000 lines, the counter emf would rise and prevent the required 14 amperes in the armature unless we received a further reduction in speed. The motor requires 14 amperes to develop the required torque, so the speed continues to fall (Figure 19-6C). Thus 1,400,000 lines instead of the 1,000,000 lines would be required in a series motor at double torque, so an inverse proportion is used: 1,400,000 : 1,000,000 = 1340 : x and we get about 957 rpm, as illustrated in Figure 19-6C.

In the results for double torque, the shunt motor drops 1 percent in speed and the series motor drops 28.6 percent in speed. Thus, in Figure 19-6C line CB no longer can represent both speed and counter emf. The counter emf is still 86 volts, but the speed is 957 on line CD.

The figures used here to illustrate the better speed regulation of a shunt motor over a series motor are quite representative, but in no way refer to any particular motor. They are used to show the difference in speed regulation between the two types of motors and have purposely been exaggerated.

It would be good to mention that, in purchasing DC motors, there is a difference between generator-supplied motors and rectifier-supplied motors. This results from the difference in AC ripple component and will have between a 1 percent and 4 percent difference in the amount of ripple.

A series motor should never be started without a load, because it tends to run away without a load and could destroy itself. The reason for this is that without load the armature theoretically requires no current. It will therefore rise in speed in an effort to generate a counter emf equal to the impressed emf.

A series motor varies most widely in speed under different loads but develops a large torque. Series motors are quite adaptable for speed control.

A compound motor varies in speed from no load to full load, but not as much as a series motor. Compound-wound motors may be differentially wound to compensate for speed variations. The terms used are *accumulative compounded* and *differentially compounded.* The design takes care of the speed variations, and this is to be taken into consideration when purchasing motors. The torque of a compound-wound motor is less than that of a series motor.

A shunt motor in most cases has the most constant speed of the three types, but develops a torque somewhat less than that of a compound-wound motor and much less than that of a series-wound motor.

Compound-wound motors are used quite often on large elevators. Since DC motors and generators are quite similar, refer to Figure 18-9 for a schematic of a series motor, Figure 18-10 for a schematic of a shunt motor, and Figure 18-11 for a schematic of a compound-wound motor, in each case replacing the load by a DC input. Figure 18-8 illustrates a shunt-wound generator with commutating poles.

It would be good to caution that under no-load, the shunt field shouldn't be opened, as the motor would tend to run away and destroy the motor. The starters for DC motors should have controls that will prevent the operation of the motor under these conditions.

Starting the DC Motor

It is good to study manufacturer's schematics of starters for DC motors and become familiar with them. Starters of all types are available commercially, and the schematics are also available from the manufacturers.

On a shunt motor, the field winding is thrown across the line input and a rheostat connected in series with the armature. At start, the rheostat resistance is in the circuit and gradually cut out of circuit as the motor gets up to speed. There are, of course, variations, depending upon the type of service for which the motor is being used.

On a series motor starter, the rheostat is in series with the field and armature.

Reversing Motor Direction

To reverse the direction of a DC motor, the field current *or* the armature current is reversed, but not both. Usually the armature current is reversed, as this is much easier to accomplish. If the motor has commutating poles, the current must also be reversed in the commutating poles as well as in the armature. Figure 19-7A and

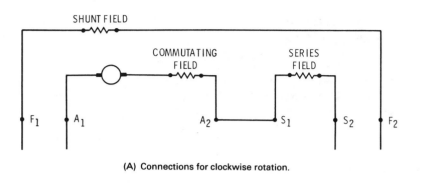

(A) Connections for clockwise rotation.

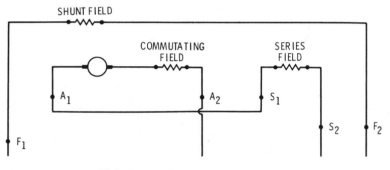

(B) Connections for counterclockwise rotation.

Figure 19-7 Rotation of compound-wound DC motor.

19-7B illustrate connections for reversing a compound-wound motor with commutating poles. The leads shown are in the connection box on the motor.

Questions

1. May a DC generator be used as a motor?

2. May a DC motor be used as a generator?

3. Where does the motor effect come from? Explain fully with sketches.

4. What is impressed emf?

5. What is effective emf?

6. What is counter emf?

7. Explain the regulation of a shunt motor under different loads.

8. Explain the regulation of a series motor under different loads.
9. What are series motors used for?
10. What are shunt motors used for?
11. What are compound motors used for?
12. Is it okay to start a series motor under no load? Explain.
13. Explain how a series motor is started.
14. Explain how a shunt motor is started.

Chapter 20

Inductance and Reactance

The subject matter is now concerned with the properties of alternating currents. Ohm's law was covered earlier in the book and will be applicable to AC circuits that have resistive loads. Many AC calculations will require Ohm's law in a little different form; it may be called the AC Ohm's law. In run-of-the-mill work, the DC Ohm's law is not used too often, but every electrician should be quite familiar with it, even though he is not in design work. In our everyday work we run into the AC Ohm's law constantly and should take every step to gain a good working knowledge of the principle involved and how to use it.

Insofar as is possible, every effort will be made to keep the material given here on a practical level that every electrician might readily understand and be able to put to everyday use.

Inductance

Inductance opposes the change of the current in the circuit and is sometimes referred to as electrical inertia. Coils are an example, but even a two-wire circuit, especially when in metallic raceways, has inductance. Every electrical circuit that forms a loop has some inductance. Inductance is an inherent property that doesn't oppose the flow of a steady current, such as DC, but it opposes any change in value of the current. It may be stated that it is a form of resistance and is especially an inherent property of coils. A straight wire of 100 feet possesses virtually no inductance. If the wire is made into a coil, inductance increases rapidly. It is a physical property and has no reference to power, current, or voltage. You may refer back to Chapter 16 on electromagnetic induction.

The unit of inductance is the *henry* (H). A coil possesses an inductance of 1 henry when a current varying at the rate of 1 ampere per second induces 1 volt in the coil.

Inductive Reactance

Inductive reactance is a direct result of inductance and is expressed in ohms. Reactive ohms are similar to resistive ohms in their effect on current limitations, but it will be found that there is an angular difference between the two. The symbol of inductive reactance is

X_L. The product of inductance (L), the frequency (f), and a constant (2π) is equal to the inductance reactance in ohms:

$$X_L = 2\pi\, fL$$

where

X_L = inductive reactance in ohms
f = frequency in hertz
L = inductance in henrys

Impedance

Theoretically, it is impossible to have a circuit with inductive reactance and no resistance. Mathematically, a particular type of function is required to relate voltage and current in an AC circuit. The one generally employed is called the *impedance* of the circuit and the symbol used is Z. The combined impedance of inductive reactance (X_L) and resistance in ohms (R) is termed impedance and is expressed by the symbol Z. Thus,

$$Z = \sqrt{R^2 + X_L^2}$$

where

Z = impedance in ohms
R = resistance in ohms
X_L = inductive reactance in ohms

If an AC circuit has resistance only, the current and voltage will be represented by sine waves in phase as illustrated in Figure 20-1.

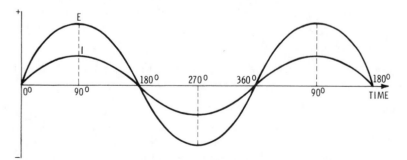

Figure 20-1 Current and voltage sine waves in an entirely resistive circuit.

If an alternator supplies an inductive load such as illustrated in Figure 20-2, the circuit will have inductance and resistance, but for this discussion the resistance will be ignored.

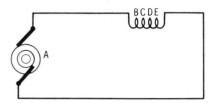

Figure 20-2 Alternator supplying an inductive load.

The changes of values and directions of currents in turn produce very important effects. They give AC its most valuable properties and produce many curious complications.

Effects of Circuit Inductance

Considering Figure 20-2, when an AC voltage is applied to coil *BCDE* from alternator *A*, the current starts to rise simultaneously in coil *BCDE*. Accompanying this current are loops of magnetic force that cut across the other turns, and Lenz's law says an emf is induced that opposes the rising current.

When the current finally reaches the maximum point, it is stationary for a moment, at which time the induced emf ceases. Then the current starts to fall off and again induces an emf in the coil that opposes this dropping current. This applied voltage has to overcome any resistance plus the opposing induced emf that is generated.

A portion of the applied voltage is used up in order to overcome the induced emf, and the remainder of the applied voltage is the resistive voltage used to force the current through the resistance.

The applied voltage and current won't give the results of Ohm's law $I = E/R$, but result in $I = E/R$ less a quantity dependent upon the amount of inductive emf encountered. In the chapter on DC motors, the applied voltage and the counter emf and effective voltage in DC motors were covered. This is similar to a motor, but in a motor the coils move and the flux is stationary, while here the coils are stationary and the flux moves.

The induced emf also takes the form of a sine wave. The results are twofold in the circuit illustrated in Figure 20-2.

1. The current in the coil *BCDE* is less than the applied voltage would indicate from Ohm's law (DC).

2. The current reaches its maximum later than the applied voltage reaches its maximum.

This results in sine waves as shown in Figure 20-3. The buildup of the current is damped by the inductance and is caused to lag behind the applied voltage. When looking at Figures 20-1 and 20-3, note that time moves to the right. In Figure 20-3 the distance AB on the zero line represents the angle (in electrical degrees) by which the current lags the applied voltage. The maximum current value A reaches a peak a little later than the applied voltage peak B. The current at D is less than its maximum value when the applied voltage is maximum at B.

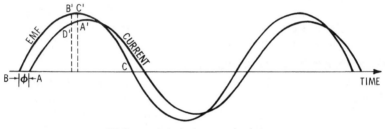

(A) Current lags the impressed voltage.

(B) Phase angle for small X_L.

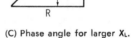

(C) Phase angle for larger X_L.

Figure 20-3 Characteristics of an AC circuit with inductance.

Phase Angle

The angle between A and B on the time and angle line is called the *phase angle* and is represented by the Greek letter ϕ. This angle of lag, ϕ, varies with the inductance in the AC circuit. It becomes greater with more inductance. In Figure 20-3B is one angle (ϕ_1) and in Figure 20-3C the inductance X_L increases, so angle ϕ_1 increases to become ϕ_2.

Assume a circuit without any resistance and only pure inductance. The phase angle ϕ between the applied voltage and current would be 90°. Looking at Figure 20-4, it may be noticed that when

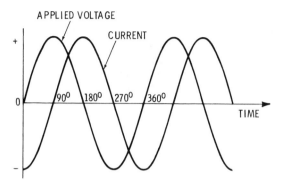

APPLIED VOLTAGE

CURRENT

TIME

Figure 20-4
Current 90° out
of phase with the
impressed emf.

the applied voltage is maximum at 90°, the current is zero, and at 270° the voltage is maximum in the negative direction and the current is again zero.

In an inductive circuit there are three voltages to be considered:

1. The applied voltage on the circuit

2. The induced emf, which again involves Lenz's law

3. The resistive voltage, which is the voltage developed across the series resistance of the inductor

The applied voltage is divided into the inductive voltage (voltage across the inductance) and resistive voltage (voltage across the resistance), as in Figure 20-5. The applied voltage is equal to the vector sum of inductive and resistive voltages. Note in Figure 20-5 that the inductive voltage leads the applied voltage, which leads the resistive voltage by the angle ϕ. The resistive voltage is in phase with the current and lags 90° behind the inductive voltage.

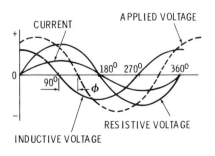

CURRENT

APPLIED VOLTAGE

ϕ

RESISTIVE VOLTAGE

INDUCTIVE VOLTAGE

Figure 20-5   Relationship of
applied voltage, inductive voltage,
resistive voltage, and current.

This may all be illustrated vectorially, as in Figure 20-6. Line AC is the applied voltage (1000 volts). The resistive voltage lags the inductive voltage by 90°, and the applied voltage leads the resistive voltage by ϕ.

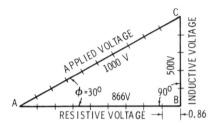

Figure 20-6 Vector representation of voltages in an AC circuit.

The easiest and proper method of calculating the resistive voltage and induced emf is by trigonometry, but it may also be done reasonably accurately by laying out vectors by scale and angles. We have a 90° angle, a 30° angle, and the hypotenuse of a right triangle with which to work, and arrive at 866 volts' resistive voltage and 500 volts' induced voltage.

By trigonometry, $AB/AC = \cos\phi$. Therefore, $AB = AC \times \cos\phi$; $AB = 1000 \times 0.866$ (volts resistive). Also, $BC/AC = \sin\phi$. Therefore, $BC = AC \times \sin\phi$ $BC = 1000 \times 0.5 = 500$ (volts induced).

$I \times Z =$ Applied Voltage
$I \times R =$ Resistive Voltage
$I \times X_L =$ Inductive Voltage

where

$I =$ current in amperes
$Z =$ impedance in ohms
$R =$ resistance in ohms
$X_L =$ inductive reactance in ohms

From these formulas, it is plain to see that with current I constant, the impedance (Z) relates to the applied voltage, the resistance (R) relates to the resistive voltage, and the inductive reactance (X_L) relates to the inductive voltage.

Figure 20-7 illustrates this statement. The vector used has the same angles as were used in Figure 20-6.

If the AC circuit has inductance, we use $I = E/Z$ instead of $I = E/R$, where Z is composed of both resistance and inductive reactance.

$$I = \frac{R}{Z} = \frac{E}{\sqrt{R_2 + X_L^2}}$$

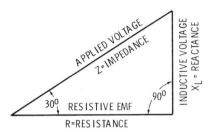

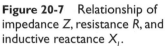

Figure 20-7 Relationship of impedance Z, resistance R, and inductive reactance X_L.

From the foregoing discussion, it is plain to see that there is a difference between applied voltage and resistive voltage, resistive voltage being the lesser of the two. Current times applied voltage won't give *effective* or *true power*, but the true power may be found by the use of cos ϕ (Figure 20-6), so P in watts = $I \times$ applied voltage $\times$ cos ϕ = $I \times$ resistive voltage.

True Power and Power Factor

Wattmeters read only true power. The applied voltage times current is the *apparent power* and both of these may be read by separate meters, a voltmeter and an ammeter. Using Figure 20-8 in this discussion, there is an inductive load X_L supplied from the AC source of power G. Ammeter A and voltmeter V register 10 amperes and 100 volts, respectively, and wattmeter W registers 900 watts of true power:

$$\frac{\text{Watts}}{\text{Volts} \times \text{Amperes}} = \frac{\text{True Power}}{\text{Apparent Power}} = \cos \phi$$

So $P/EI = 900/(100 \times 10) = 0.90 = \cos \phi$, and thus $0.90 \times 100 = 90$ percentage. The cosine of the angle is called the *power factor*.

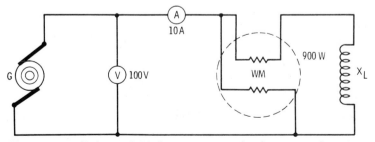

Figure 20-8 Relationship of wattmeter and voltmeter-and-ammeter readings.

Instead of a decimal, which is cos ϕ, use cos $\phi \times 100$, which will give the answer in percent.

Power factor may be either leading or lagging. With inductive reactance X_L, the power factor is lagging. This is to say, the current in the circuit lags behind the applied voltage. Leading power factor will be discussed in a later chapter, covering capacitive reactance.

When power factor is involved, the circuitry must be sized to supply the apparent power, while actual work done is accomplished by the true power. More will be covered on this in a later chapter.

In dealing with AC circuitry, the terms volt-amperes (VA) and kilovolt-amperes (kVA) will be often used. VA is the applied voltage times amperes, and kVA is VA/1000. With 100 percent power factor (PF), VA will be the same as watts (W), and kVA will be the same as kilowatts (kW). A 100 percent power factor is termed *unity power factor*. With other than unity power factor, VA $\times$ PF = watts, and kVA $\times$ PF = kW.

Worked Examples

Theory in itself is a necessity, but one must also be able to apply this theory in practice. Thus, problems will be given:

1. Find the inductance of a coil that has an inductive reactance of 1.884 ohms at 60 Hz.

   Now $X_L = 2\pi\ fL$. Transposing,

   $$L = X_L/6.25\ f = 1.884/(6.28 \times 60) = 5\ \text{mH}$$

2. A 20-henry filter choke is used on a 150-volt, 60-Hz power supply.

   (a) Find the inductive reactance in ohms.

   (b) What current will flow when the choke is across the line?

   (a) $X_L = 2\pi\ fL = 6.28 \times 60 \times 20 = 7536\ \text{ohms}$

   (b) $I = E/2\ X_L = 150/7536 = 0.02\ \text{amperes}$

3. Recall that $Z = \sqrt{R^2 + X_L^2}$. A coil has 5 ohms' resistance and 12 ohms' inductive reactance. What is the impedance of the coil?

   $$Z = \sqrt{5^2 + 12^2} = \sqrt{25 + 144} = 13\ \text{ohms}$$

4. In Figure 20-9, find the total impedance and the total current flowing.

$$Z = \sqrt{(5 + 6)^2 + (10 + 12)^2} = \sqrt{11^2 + 22^2} = 24.6 \text{ ohms}$$

$$I = 100/24.6 = 4 \text{ amperes}$$

With resistance and/or inductive reactance in parallel, the reciprocal of the reciprocals will be used as with DC Ohm's law.

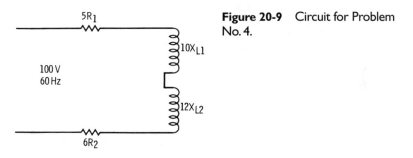

Figure 20-9 Circuit for Problem No. 4.

5. Two impedances A and B as in Figure 20-10 are connected in parallel on a 440-volt circuit at 60 Hz. Impedance A has a resistance of 50 ohms and an inductance of 0.02 henry, and B has a resistance of 80 ohms and an inductance of 0.06 henry.

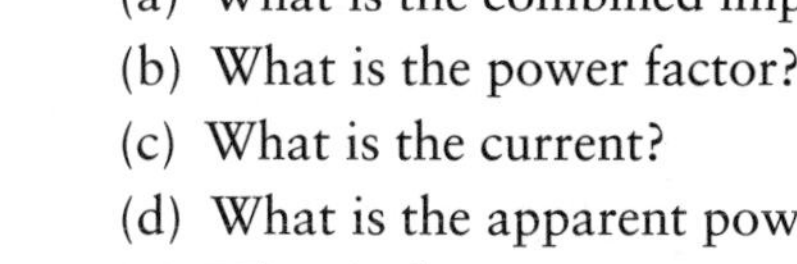

Figure 20-10 Circuit for Problem No. 5.

(a) What is the combined impedance of A and B?
(b) What is the power factor?
(c) What is the current?
(d) What is the apparent power?
(e) What is the true power?

Now $X_L = 2\pi fL$, so

$$X_{L(A)} = 6.28 \times 60 \times 0.02 = 7.54 \text{ ohms}$$

$$X_{L(B)} = 6.28 \times 60 \times 0.06 = 22.61 \text{ ohms}$$

$$Z_b = \sqrt{R^2 + X_{L(A)}^2} = \sqrt{50^2 + 7.54^2} = 50.56 \text{ ohms}$$

$$Z_B = \sqrt{R_A^2 + X_{L(A)}^2} = \sqrt{80^2 + 22.6^2} = 83.1 \text{ ohms}$$

$$I_A = \frac{E}{Z_A} = \frac{440}{50.56} = 8.7 \text{ amperes}$$

$$I_B = \frac{E}{Z_B} = \frac{440}{83.1} = 5.3 \text{ amperes}$$

$$I = 8.7 + 5.3 = 14 \text{ amperes} \hfill \text{[Ans. (c)]}$$

$$Z_r = \frac{E}{I_r} = \frac{440}{14} = 31.4 \text{ ohms} \hfill \text{[Ans. (a)]}$$

$$Z_r = 1/\left(\frac{1}{Z_A} + \frac{1}{Z_B}\right) = 1/\left(\frac{1}{50.56} + \frac{1}{83.1}\right) = 31.25 \text{ ohms}$$

$$R_r = 1/\left(\frac{1}{50} + \frac{1}{80}\right) = 30.77 \text{ ohms}$$

$$\text{PF} = \frac{R}{Z} = \frac{30.77}{31.25} = 0.984 \text{ and } 0.984 \times 100 = 98.4 \hfill \text{[Ans. (b)]}$$

$$\text{VA} = I_r \times E_r = 14 + 440 = 6160 \hfill \text{[Ans. (d)]}$$

$$\text{Watts} = \text{VA} \times \text{PF} = 6160 \times 0.984 = 6061 \text{ watts} \hfill \text{[Ans. (e)]}$$

Formulas

$$X_L = 2\pi fL = 6.28 fL$$

$$AB/AC = \cos \phi$$

$$AB = AC \cos \phi$$

$$BC/AC = \sin \phi$$

$BC = AC \sin \phi$

$I \times Z = $ Applied Voltage

$I \times R = $ Resistive Voltage

$I \times X_L = $ Inductive emf

$$I = \frac{E}{Z} = \frac{E}{\sqrt{R^2 + X_L^2}}$$

$$\frac{\text{Watts}}{\text{Volt} - \text{Amperes}} = \frac{\text{Real Power}}{\text{Apparent Power}} = \cos \phi$$

$P_{in}(\text{watts}) = \text{VA} \times \text{PF}$

$P_{in}(\text{kilowatts}) = \text{kVA} \times \text{PF}$

$$Z_r = \frac{1}{\dfrac{1}{Z_1} + \dfrac{1}{Z_2} + \cdots + \dfrac{1}{Z_n}}$$

$\text{PF} = R/Z$

Questions

1. Explain what inductance is.
2. Explain what inductive reactance is.
3. Give the formula for inductive reactance.
4. Explain the effect of induced emf.
5. Explain fully what impedance is.
6. In a circuit with inductive reactance, does the current lead or lag?
7. $I \times Z = $?
8. $I \times R = $?
9. $I \times X_L = $?
10. Give the formula for Z.

11. Wattmeters read power. (True or false?)

12. The voltmeter reading times the ammeter reading gives power. (True or false?)

13. Real Power/Apparent Power = ?

14. Explain power factor fully.

15. Does inductive reactance cause a leading or lagging power factor?

16. Three impedances, A, B, C, are connected in parallel in a 60-Hz circuit. Impedance A has 200 ohms' resistance. Impedance B has 60 ohms' resistance and an inductance of 0.5 henry. (a) What is the combined impedance? (b) What is the power factor? (c) What is the current in each branch at 120 volts? (d) What is the combined current in all three branches at 120 volts?

17. A coil possessing 10 ohms' resistance and 8 ohms' inductive reactance is connected in series with a coil of 25 ohms' resistance and 12 ohms' inductive reactance. What voltage will be required to force 5 amperes through this circuit?

Chapter 21

Capacitance in AC Circuits

Capacitors and capacitance were thoroughly covered in theory in an earlier chapter. These both play a very important part in AC circuits and calculations.

Analogy of Capacitor Action

Figure 21-1 is an analogy of how AC apparently flows through a capacitor. In reality it doesn't flow through the capacitor, but the results are very similar to its flowing through the capacitor. A rubber diaphragm (D) forms a tight seal and may be compared to the dielectric of a capacitor. The upper and lower halves of the enclosure (C) may be compared to the two plates of a capacitor. Regions X and Y may be compared to conductors connected to the two plates of the capacitor, and plunger P may be compared to an AC source.

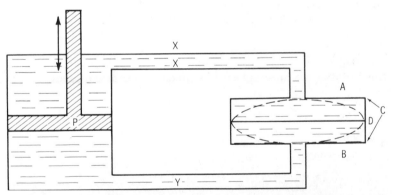

Figure 21-1 Analogy illustrating how AC can apparently flow through a capacitor.

When plunger P goes down, fluid moves in Y toward diaphragm D and the diaphragm is forced into position A. This may be compared to a current charge in a capacitor. Plunger P then moves up and the fluid in X flows toward diaphragm D from the underside of D back through Y to the plunger, and diaphragm D goes to position B.

A capacitor reverses plate charges every alternation in an AC circuit.

Capacitor Action

Beginning with a simple circuit, as in Figure 21-2, will aid in a better understanding of cause and effect. Here there is an ammeter, A, two capacitor plates, C and D, dielectric K, battery B, and switch S.

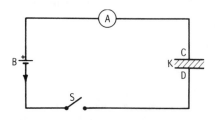

Figure 21-2 Current surge into a capacitor.

Switch S is open, plates C and D possess no electrical charge, and ammeter A reads zero. At the moment switch S is closed there is an inrush of current from battery B into plate D, charging it negative. Electrons from plate C go back through ammeter A to battery B. The maximum current results at the instant the switch S is closed and tapers off to zero as the plates become charged. The major point here is that there is maximum current when there is no charge in the capacitor.

Capacitance

The capacitance of a capacitor is fixed by its construction and is proportional to three things:

1. Size and shape of the plates
2. Thickness of the dielectric
3. Material of the dielectric

It will be recalled that a capacitor has a capacitance of 1 farad, when 1 ampere flowing for 1 second raises its potential 1 volt.

$$C = \frac{Q}{E}$$

where

Q = charge in one plate in coulombs
E = voltage applied across capacitor in volts
C = capacitance in farads

Current and Voltage Relations

In Figure 21-3, the relationship of current and applied voltage, as they appear in a capacitor, will be shown. You will recall that when the voltage was first applied to a capacitor, the current was its greatest. This was illustrated in Figure 21-2. Now, referring to Figure 21-3, the applied voltage starts at zero and moves in time from 0 to B, and in voltage from 0 to C. The voltage rises rapidly and it is found that the current during this interval, DE, is near maximum rate. At point F the voltage is at maximum, so at this point (F) there is no voltage change and no current (G), but the quantity of charge in the capacitor is maximum. Here, since the capacitor is fully charged, there is no current. Also, we see that the voltage of the capacitor is maximum at M and in opposite polarity to the applied voltage. (The voltage across the capacitor acts so as to buck the applied voltage.) When the applied voltage, F to T, falls off, the voltage across the capacitor will cause a reversal of current from G to N.

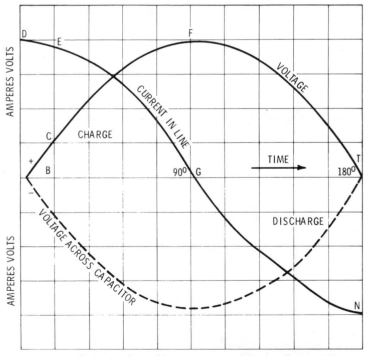

Figure 21-3   Relationship of line current, applied voltage, and counter emf in a circuit with capacitive reactance.

In one voltage cycle the capacitor is charged and discharged twice. The time for each charge and discharge is $t = \frac{1}{4}f$, where t is the time in seconds and f is the frequency of the applied voltage in hertz.

Due to the change of value of this current from maximum to zero, each quarter-hertz of the total charge Q that will flow in or out of the capacitor acts as though it were flowing through the capacitor and is equal to the rate of flow (I) multiplied by the time (t) of flow: $Q = $ average $I \times t$.

Capacitive Reactance

Inductance in an AC circuit causes inductive reactance, as covered in the last chapter, and is represented by X_L. Capacitance in an AC circuit causes capacitive reactance and is represented by X_C. Both inductive reactance and capacitive reactance are forms of AC resistance. They have opposite effects and one tends to cancel out the other, as will be covered later.

$Q = I_{avg} \times (\frac{1}{4}f) = I_{avg}/4\,f$. Then, combining $Q = CE$ and the above, we have $CE_{max} = I_{avg}/4\,f$, so that $E_{max} = I_{avg}/4\,fC$; since $I_{avg} = 0.636\,I_{max}$, $E_{max} = 0.636\,I_{max}/4\,fC = I_{max}/6.28\,fC$. Dividing both sides by I_{max} gives a ratio that is expressed in ohms:

$$\frac{E_{max}}{I_{max}} = \frac{I_{max}/6.28\,fC}{I_{max}} = \frac{1}{6.28\,fC}$$

which is the expression for capacitive reactance X_C.

A farad is not a very practical unit, so we find microfarads used, which is 1/1,000,000 of a farad, or 10^{-6} farad.

For 60 Hz, the formula $1/6.28\,fC$ or $1/2\pi\,fC = X_C$ may be written $X_C = 1/(2 \times 3.14 \times 60 \times C) = (1/377C) = $ or $0.00265/C$ in farads or $2650/C$ in microfarads.

As an example, use Figure 21-4 and find the capacitive reactance and current:

$$X_c = \frac{2650}{2} = 1325 \text{ ohms}$$

$$I = \frac{E}{X_c} = \frac{2000}{1325} = 1.5 + \text{amperes}$$

Inductive and Capacitive Circuits

From the discussion of Figure 21-3, it may be seen that capacitive currents lead the impressed emf by 90°, which is also shown in

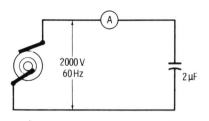

Figure 21-4 Finding current and capacitive reactance.

Figure 21-5A. In review, compare Figure 21-5A with the impressed emf and current in an inductive circuit (Figure 21-6), where the current lags the impressed emf by 90°.

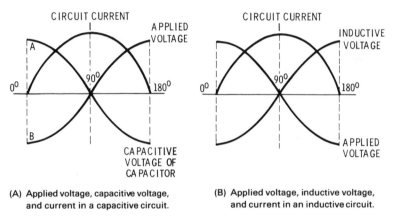

(A) Applied voltage, capacitive voltage, and current in a capacitive circuit.

(B) Applied voltage, inductive voltage, and current in an inductive circuit.

Figure 21-5 Relationships in reactive circuits.

From this, three important statements may be made:

1. The presence of inductance in a circuit causes the current to be damped or diminished in amount and to lag in phase.

2. The presence of capacitance in a circuit causes the current to increase in amount and to lead in phase.

3. Thus, when both are present, one will always tend to neutralize or offset the effect of the other.

Either inductance or capacitance in a circuit by itself is objectionable and causes poor power factor. Inductance causes a lagging power factor and capacitance causes a leading power factor. Either one by itself is objectionable. If either one or the other is present in a circuit, the addition of the other will improve the power factor.

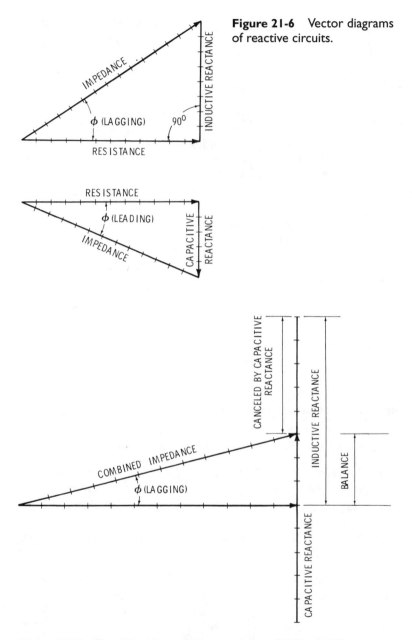

Figure 21-6 Vector diagrams of reactive circuits.

Figure 21-7 Combined impedance: resultant of R, X_L, and X_C.

In the last chapter it was stated that $I = E/Z$, Z being impedance. Also learned was that $Z = \sqrt{(R^2 + X_L^2)}$, for inductive circuits. Now, for capacitive circuits, $Z = \sqrt{(R^2 + X_C^2)}$.

A vector analysis of this will help. In Figure 21-6A is shown a vector diagram for an inductive circuit. Figure 21-6B is a vector diagram for a capacitive circuit set up on the same units of measurement. Then, Figure 21-7 is a vector diagram of the combination of vectors in Figure 21-6A and 21-6B, showing the reduced impedance and the reduced angle ϕ or power factor (PF). These vectors are rather self-explanatory. The inductive reactance is plotted up and the capacitive reactance is plotted down, as they are in opposition to each other. Thus, when plotted together in Figure 21-7, the capacitive reactance, being the smaller, is plotted up above the resistance, so it may be deducted from the inductive reactance, leaving a smaller inductive reactance than we started with, and angle ϕ is reduced; thus the power factor is reduced.

Formulas

$$C = Q/E$$

$$t = 1/4\,f$$

$$Q = I_{avg} \times t$$

$$X_c = \frac{1}{6.28\,fC} \quad \text{or} \quad \frac{1}{2\pi\,fC}$$

$$X_c = \frac{1}{377C} \quad \text{for} \quad 60\ \text{Hz}$$

$$V_c = \frac{0.00265}{C\ (\text{in farads})} \quad \text{for} \quad 60\ \text{Hz}$$

$$X_c = \frac{2650}{C\ (\text{in}\ \mu F)}$$

$$Z = \sqrt{R^2 + X_c^2}$$

Questions

1. Give a hydraulic analogy to a capacitor connected to AC.
2. When an emf is applied to a capacitor, at what point is the maximum current reached?

3. On what does the capacitance of a capacitor depend?
4. Does the current lead or lag the impressed emf in a capacitive circuit, and what will the angle ϕ be in degrees?
5. What is the difference between farads and microfarads?
6. Give the formula for capacitive reactance.
7. Give the formula for X_C when 60 Hz and microfarads are involved.
8. If capacitance is added to an inductive circuit, explain what happens to the power factor.
9. When capacitance is added to an inductive circuit, what happens to the circuit impedance?
10. Give the formulas for impedance when resistance and capacitance are involved.

Chapter 22

Resistance, Capacitance, and Inductance in Series and Parallel

This chapter will cover calculations of resistance, capacitance, and inductance in series and parallel.

RC Circuit

Figure 22-1 shows a 100-ohm noninductive resistor in series with a 4-μF capacitor, connected to a 130-Hz AC source of 2000 volts.

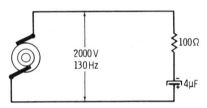

Figure 22-1   Resistance and capacitance in series.

Let us find,

1. Capacitive reactance in the circuit
2. Total impedance of the load
3. Current in the circuit
4. Power factor of the circuit
5. Angle of lead in degrees

$$X_C = \frac{1}{2\pi f C} = \frac{1}{6.28 \times 130 \times 0.000004} = 306 \text{ ohms} \quad (1)$$

$$Z = \sqrt{R^2 + X_C^2} = \sqrt{100^2 + 306^2} = 322 \text{ ohms} \quad (2)$$

$$I = E/Z - 200/322 = 6.21 \text{ amperes} \quad (3)$$

$$\cos\phi = R/Z = 100/322 = 0.311$$

$$PF = \cos\phi \times 100 = 0.311 \times 100 = 31.1\% \quad (4)$$

From the table of cosines at the back of this book,
$$0.311 = 72° \quad (5)$$

This may be plotted vectorially. See Figure 22-2. Draw a line, *AB*, 100 units long for *R*. Then at 90° from *AB*, draw *BC* 306 units long for X_C. Connect *A* and *C* by a line, measure the units of length in this line, and it will be found to be 322 units in length or 322 ohms impedance.

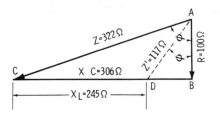

Figure 22-2   Vector value of *R*; X_C and X_L in series as shown in Figure 22-1.

With a protractor, measure angle ϕ and it will be found to be 72°. Since this is a capacitive circuit, the power factor will be leading and PF = $(R/Z) \times 100 = (100/322) \times 100 = 31.1\%$.

RLC Circuit

Next add a coil into the circuit as shown in Figure 22-3. This coil is 0.3 H.

1. Find the inductive reactance of the coil.

2. Find the total impedance of the circuit.

3. Find the current in the circuit.

4. Find the power factor.

5. Find the angle ϕ in degrees.

$$X_L = 6.28fL = 6.28 \times 130 \times 0.3 = 245 \text{ ohms} \tag{6}$$

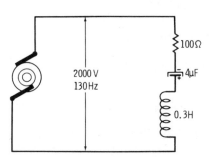

Figure 22-3   Inductance added to the circuit of Figure 22-1.

$$Z = \sqrt{R^2 + (X_C - X_L)^2} = \sqrt{100^2 + (306 - 245)^2}$$

$$= \sqrt{100^2 + 61^2} = 117 \text{ ohms} \tag{7}$$

$$I = E/Z = 2000/117 = 17.09 \text{ amperes} \tag{8}$$

$$\cos\phi = R/Z = 100/117 = 0.855$$

$$PF = \cos\phi \times 100 = 0.855 \times 100 = 85.5\% \tag{9}$$

Cosine ϕ is 0.855, and from the table is 31°. $\tag{10}$

The capacitive reactance is greater than the inductive reactance, so the power factor is leading.

Now in Figure 22-3 measure CD up from C as 245 units of length (inductive reactance), which has to be subtracted from the capacitive reactance, BC. This leaves the effective capacitive reactance as BD, or 61 ohms X_C. Angle ϕ' can be measured by a protractor as 31° leading.

No circuit actually can have inductance without some resistance. So there are core losses and copper losses. With a capacitor there is a loss required to reverse the dielectric strain. The two problems just covered therefore did not include these losses. The values for a problem covering such losses are illustrated in Figure 22-4.

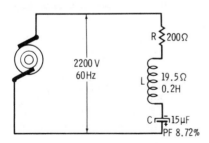

Figure 22-4   Resistance, inductive reactance, and capacitive reactance in series.

Find:

1. Total current

2. PF of the entire circuit

In conjunction with this problem, Figure 22-5 will be used. Line AB, 200 units long, represents the resistance of R. The inductive

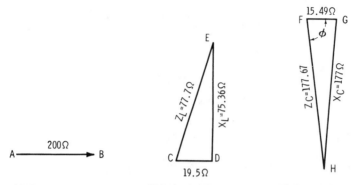

(A) Resistor vector.   (B) Inductor vectors.   (C) Capacitor vectors.

Figure 22-5   Vector diagram for impedance of R, L, and C of Figure 22-4.

reactance of coil L is X_L = 6.28 × 60 × 0.2 = 75.36 ohms. The impedance of coil L is

$$Z_L = \sqrt{R_L^2 + X_L^2} = \sqrt{19.5^2 + 75.36^2} = 77.8 \text{ ohms}$$

The impedance vector for inductor L is given by CE. The capacitive reactance of C is

$$X_C = \frac{1}{6.28fC} = \frac{1}{6.28 \times 60 \times 0.000015} = 177 \text{ ohms}$$

There is no ohmic resistance in a capacitor, so the energy component of a capacitor can't be expressed as such. It is, however, possible to express the energy component in equivalent ohms of the total impedance of the capacitor. To do this, the power factor of the capacitor must be obtained. This may be done with a voltmeter, ammeter, and a wattmeter as covered in Chapter 20: W/VA = cos ϕ, and cos ϕ × 100 = PF.

From these figures, vector diagram FGH may be constructed. Line GH was found to be 177 ohms. If the power factor is known, the energy component, FG, and the total impedance may be found. Now $FH = GH/\sin \phi_2$.

Since neither angle ϕ_2 nor its sine are known, this can't be found directly. The power factor of the capacitor is cos ϕ_C, so if PF = 0.0872 = cos ϕ_C, the sine of ϕ_C may be found by using a table.

Then, applying the above formula, $FH = GH/\sin \phi_C = 177/0.9962 = 177.67$ or Z_C. $Z_C \times \cos \phi_C = 177.67 \times 0.0872 = 15.49$, which is line FG or the energy component of the capacitor.

To get the combined impedance of these three devices,

$$R = 200 + 15.49 + 19.5 = 234.99 \text{ ohms}$$

$$Z = \sqrt{R^2 + (X_C - X_L)^2} = \sqrt{234.99^2 + (177 - 75.36)^2}$$

$$= 256.22 \text{ ohms.}$$

The current is $I = E/Z = 2200/256.2 = 8.5$ amperes.

$$PF = (R/Z) \times 100 = 234.99/256.22 \times 100 = 91.7\%$$

Figure 22-6 is the combined vectors shown in Figure 22-5 for the impedance of R, L, and C in Figure 22-4. With vectors, the wattless

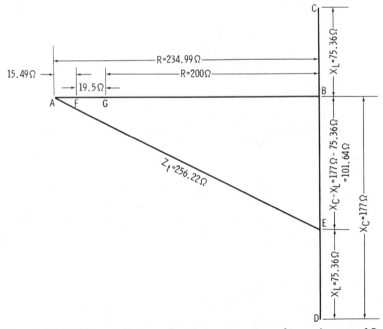

Figure 22-6   Vector diagram showing summation of impedances of R, L, and C in Figure 22-4.

component of capacitors is vertical and below the true component, e.g., as *BD* is the wattless component of the capacitor. Line *AF* is the true-power component of the capacitor, *FG* the true component of the inductance, and *GB* is the resistance of *R*. The wattless component of inductance is vertical (*BC*) and above the true component, *FG*.

The common expressions wattless power or wattless current are used with AC. There is actually no wattless current, since wherever current flows there must be an expenditure of energy. These currents are out of phase with the applied voltage, and the consideration of power involved is based on energy components and wattless components of these currents. Figure 22-7 illustrates the relationship of the three components in an inductive circuit.

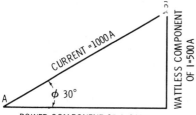

POWER COMPONENT OF I = 866 A

Figure 22-7 Vector diagram showing typical relationships of current components in an AC circuit.

Taking the values of *I* in Figure 22-7, and using 1000 volts, the line *AB* (1000 A) times 1000 V = 1,000,000 VA apparent power, or 1000 kVA. *BC* (500 A) times 1000 V = 500,000 VA as the wattless power component, and *AC* (866A) times 1000 V = 866,000 watts or 866 kW of true power.

This tells us that the vector diagram in Figure 22-7 is an inductive circuit with an angle lag of 30° or a power factor of 86.6%. Thus, the circuit must be designed for 1000 amperes due to the 86.6% PF, while 866 amperes would be all the actual current that would be required to do the work if we had unity (100%) power factor.

Parallel LC Circuit

Inductance and capacitance are not often found in series; they are usually in parallel. There is a capacitance between conductors of circuits; the conductors are the plates and the insulation, and air the dielectric.

Take a practical example, the circuit shown in Figure 22-8. The system has a 2-μF capacitor and the load is 0.5 H, voltage 2000 V, and frequency 130 Hz.

$$X_C = \frac{1}{6.28fC} = \frac{1}{6.28 \times 130 \times 0.000002} = 613 \text{ ohms}$$

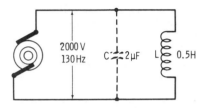

Figure 22-8 Capacitive effect of conductors.

The energy components will be neglected in this example, so C is all X_C or Z_C, and L is all X_L or Z_L. Thus,

$I_C = E/Z_C = 2000/613 = 3.26$ amperes in C (leading)

$I_L = E/X_C = 2000/408 = 4.9$ amperes in L (lagging)

I_C is thus out of phase 90°, so is all wattless component

I_L is also out of phase 90°, so is likewise wattless component

$I_L - I_C = 4.9 - 3.26 = 1.64$ amperes is the total current supplied by the alternator. Now, $Z = E/I = 2000/1.64 = 1219$ ohms combined impedance.

Since we have neglected the energy losses, the power factor is zero.

The current required is considerably less than that required by either device, due to the fact that C and L are diametrically opposed to each other. The capacitor acts as sort of a generator, storing energy in one alternation and releasing it in the next alternation.

Resonance

With capacitance and inductance in series, X_C and X_L are in series, and as $X_L = 6.28fL$ approaches the value of $X_C = 1/6.28fC$, the line voltage and current will increase. When X_C and X_L are in parallel (as in Figure 22-8), the line current diminishes as X_C approaches X_L.

Circuits possessing X_C and X_L result in resonance. Resonance can't exist without inductance and capacitance. Perfect resonance results when $X_C = X_L$.

To illustrate, the alternators in Figure 22-9 and the capacitors are such that voltage builds up as it does when X_C and X_L balance. In Figure 22-9, alternator A charges C to 2000 volts. In Figure 22-9B, the alternator has reversed polarity. Capacitor C, in series with A, sends its emf back so that at the end of that alternation the two voltages have added, and C is now charged with 4000 volts, in Figure 22-9C. In Figure 22-9D the polarity of the alternator reverses, and the action in Figure 22-9B is repeated, except the capacitor receives a charge of 6000 volts, and so on. (In reality, this circuit would have to be modified to behave as shown in Figure 22-9.)

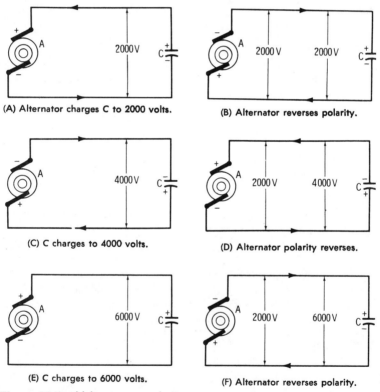

(A) Alternator charges C to 2000 volts.

(B) Alternator reverses polarity.

(C) C charges to 4000 volts.

(D) Alternator polarity reverses.

(E) C charges to 6000 volts.

(F) Alternator reverses polarity.

Figure 22-9 Voltage accumulation.

Resonance occurs at a certain frequency. Should the frequency be increased, the value of X_L will rise; but at the same time that an increase in frequency increases X_L, it reduces X_C. Where resonance exists, altering the frequency will check it. The smaller the resistance in a series circuit, the greater is the local voltage set up across the capacitor and inductance.

Voltage at resonance is very dangerous and undesirable, because of the damage and possible breakdown of insulation due to the increase in voltage. Ferroresonance in power systems must be avoided by proper designs.

Current resonance is highly desirable because it relieves the alternator of the necessity of furnishing the wattless components of the current and tends to correct the power factor toward unity, or 100%.

System Power Factor

The power factor of a system is of prime importance. A 25,000-kVA system at 80% PF would mean that the actual energy received is $25,000 \times 0.8 = 20,000$ kW true power from the system. The system equipment from alternators through transmission lines, distribution lines, etc., has to be large enough to handle the current associated with 25,000 kVA, while 20,000 kW is all that is working.

Alternators will motor, and when motoring, if the field is overexcited, the alternator will act as a capacitor, causing the current to lead the impressed voltage. Such units are used to correct power factors of highly inductive loads. These are called *synchronous* or *rotary capacitors*.

Capacitors may be installed at motors or elsewhere on a system, in parallel with the load, and thus compensate for the inductive reactance.

Utilities have power factor clauses in their energy rates to large users, that is, when the power factor drops, say, to 90%, the rate goes up. As was stated earlier, wattmeters register true power and not apparent power, so billings for electrical energy ignore the power factor involved and the utilities are entitled to some compensation for this. Power factor meters are installed to indicate or record power factor for rate determination.

Consider the parallel circuit in Figure 22-10 with values as given in the illustration. Find

1. Current in branch A
2. Current in branch B
3. Current in branch C
4. Total current
5. Impedance of branch B
6. Impedance of branch C

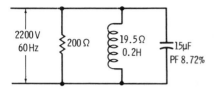

Figure 22-10 Resistance, inductance, and capacitance in parallel.

7. Combined impedances
8. Power factor

Branch A has 200 ohms resistance. Branch B has $X_L = 6.28fL = 6.28 \times 60 \times 0.2 = 75.36$ ohms inductive reactance.

$$Z_B = \sqrt{R^2 + X_L^2} = \sqrt{19.5^2 + 75.36^2} = 77.8 \text{ ohms} \qquad (11)$$

$$X_C = \frac{1{,}000{,}000}{6.28fC} = \frac{1{,}000{,}000}{6.28 \times 60 \times 15}$$

$$= 177 \text{ ohms capacitive reactance}$$

Cos ϕ for 8.72% PF = 0.0872 and from the table, sin ϕ = 0.996.

$$Z_C = \frac{X_C}{\sin \phi} = 177/0.996 = 177.71 \text{ ohms} \qquad (12)$$

Z_R = plain R, as it is a noninductive circuit

$$I_A = E/Z_R = 2200/200 = 11 \text{ amperes} \qquad (12a)$$

$$I_B = E/Z_B = 2200/77.8 = 28.3 \text{ amperes} \qquad (12b)$$

$$I_C = E/Z_C = 2200/177.71 = 12.4 \text{ amperes} \qquad (12c)$$

The power factor of branch B is

$$\cos \phi = R/X_L = 19.5/77.8 = 0.2506 \quad \text{or} \quad 24.6\%$$

The energy component of I_B is

$$I = I_B \cos \phi = 28.3 \times 0.2509 = 7.1 \text{ amperes}$$

The wattless component of I_B (from the table, sin ϕ_B = 0.966) is

$$I = I_B \sin \phi_B = 28.3 \times 0.966 = 27.3 \text{ amperes}$$

The power factor of branch C is 8.72%, so cos ϕ_C = 0.0872, and from the table, sin ϕ_C = 0.996.

The energy component of I_C is

I_C cos ϕ_C = 12.3 × 0.0876 = 1.07 amperes.

The wattless component of I_C is

I_C sin ϕ_C = 12.3 × 0.0996 = 12.25 amperes.

The total current in the circuit will be the sum of the currents in A, B, and C, taking into account the phase angles involved.

The energy components of all three branches are in phase, so they may be added. For A, this is 11 amperes; for B, this is 7.1 amperes; and for C, this is 1.07 amperes: 11 + 7.1 + 1.07 = 19.17 amperes of energy component or I_e.

The wattless current for A is zero; for B, 27.3 amperes; and for C, 12.25 amperes. The wattless current for C will be subtracted from the wattless current for B or 27.3 − 12.25 = 15.05 amperes.

Formulas

$$Z = \sqrt{R^2 + (X_L - X_C)^2}$$

$$X_L = 6.28fL$$

$$X_C = \frac{1}{6.28fC}$$

$$I_C = E/Z_C$$

$$I_L = E/X_L$$

Questions

1. A coil has 5 ohms resistance and 7.5 ohms inductive reactance. A capacitor with capacitive reactance of 20 ohms and a power factor of 0.06 is connected in parallel with the coil on a 220-volt circuit.

   (a) What is their combined impedance?

   (b) What is the current in each device?

   (c) What is the total current?

   (d) What is the overall power factor?

2. What causes resonance?

3. With capacitive reactance and inductive reactance in parallel, what happens to the line current as X_C approaches X_L?

4. With X_L and X_C in series, what happens to the line current and voltage as resonance is approached?

5. Does the frequency affect resonance?

6. Is current resonance desirable?

7. What effect does current resonance have on power factor?

8. In a 60-Hz circuit of 120 volts, the current is 12 amperes, and the current lags the voltage by 60°. Find (a) the power factor, (b) the power in volt-amperes, and (c) the power in watts.

9. Give two methods of power factor correction.

10. The cosine of ϕ is 0.866. What is the sine of ϕ?

Chapter 23

Polyphase Circuits

Thus far, only single-phase (1ϕ) AC circuits have been considered. Single-phase has a very important role in the use of electricity; by the same token, polyphase circuits are also very important.

Multiphase Systems

This discussion will primarily cover three-phase (written 3ϕ) circuits. Mention must be made that at one time it was felt that two-phase would be the most practical polyphase system. It was found, however, that three-phase was more practical and economical. The only two-phase equipment currently being made is for replacement purpose. There are also six-phase systems. Six-phase systems are sometimes used in special rectifier circuits, but the six-phase is derived from rotary converters with a three-phase input and a six-phase output or from special transformers with three-phase input and six-phase output.

Two-Phase Systems

Figure 23-1 illustrates two alternators, namely, No. 1 and No. 2. For this explanation, envision both alternators connected to the same shaft, so that their windings are 90 electrical degrees apart. A voltage of 100 volts has been selected for the output voltage of each alternator.

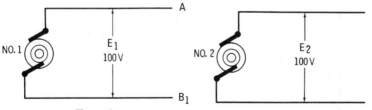

Figure 23-1   Two-phase system.

In Figure 23-2 the two alternator windings are illustrated 90° out of phase. There are two methods of two-phase transmission: One is four-wire and the other three-wire. For three-wire transmission, B_1 and B_2 are connected together, forming B for one conductor, and A and C as the other two conductors.

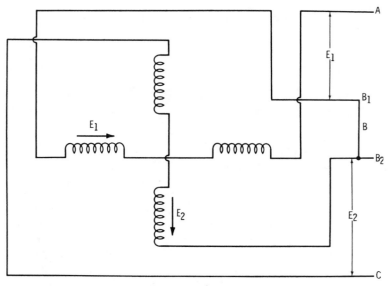

Figure 23-2 Two-phase windings 90° apart.

This three-wire connection gives 100 volts between B and C, 100 volts between A and B, and 141 volts between A and C. This is shown by the vector diagram in Figure 23-3. Here E_1 is 100 volts and E_2 is 100 volts. These two voltages are 90° out of phase, so the diagonal line E will be $100 \times \sqrt{2} = 100 \times 1.41 = 141$ volts between A and C in Figure 23-2.

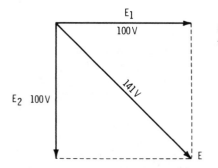

Figure 23-3 Vector diagram of two-phase voltages.

With four-wire two-phase, there would be two 100-volt circuits with no connection between them, as illustrated in Figure 23-4. Windings A and B are not connected together. Winding A supplies

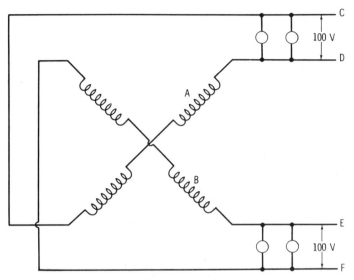

Figure 23-4 Two-phase four-wire system.

conductors *C* and *D*. Winding *B* supplies conductors *E* and *F.* The two circuits are 90° out of phase.

Three-Phase Systems

Three-phase could be considered similar to two-phase, namely, three one-phase alternators connected to one shaft so that each alternator is in turn 120° out of phase with the preceding alternator. To be more practical, a single alternator has three windings on the stator or rotor, as the case may be. These windings are connected so that they are 120° apart. They may be connected either *wye* or *delta*. Figure 23-5A illustrates a wye connection. The three phases are *A*, *B*, and *C*, with all windings connected at a common point *X*. Figure 23-5B illustrates a delta connection. The three phases, *A-B, B-C,* and *C-A,* connect as shown in this illustration.

Single-phase alternators with distributed windings will have less output than with concentrated windings. The reason for this is that part of the voltage is obtained at a disadvantage due to the various phase angles between the sections of the windings. To correct this, if the sections could be made to deliver their voltages independently, a closer approximation to the arithmetic sums of their separate voltages could be more nearly obtained.

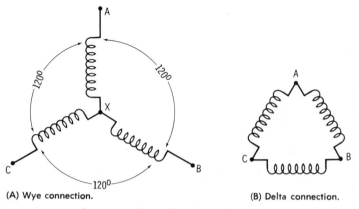

(A) Wye connection.   (B) Delta connection.

Figure 23-5 Common three-phase connections.

Figure 23-6 shows a three-phase alternator winding connected in wye as shown in Figure 23-5A. The windings are 60° apart, but if they are connected in separate circuits as shown in Figure 23-6, a great gain is affected. Conductor E is in series with conductor F, which is 180° away from E. Conductor G, 60° away from E, is in series with H, which is 60° away from F. Conductor I is in series with conductor K and is 60° away from G.

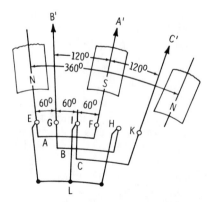

Figure 23-6 Three-phase alternator winding–connected wye.

Assuming that the six conductors were brought out to three separate circuits, a voltage output such as in Figure 23-7 would be the result. For the purpose of economy, it is desirable to combine these into one circuit, with three conductors instead of six conductors.

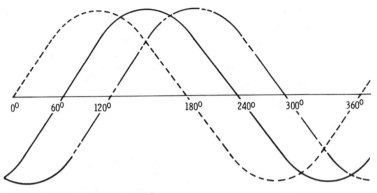

Figure 23-7 Three emfs 60° apart.

The most economical winding is the placing of conductors 60° apart, as shown in Figure 23-6.

It is then possible to connect these windings so as to deliver voltages 120° apart as shown in Figure 23-8. Points *E, I,* and *H* in Figure 23-6 are 120° apart in phase and are brought out to a common point, *L.* The other ends of *E, I,* and *H,* namely, *G, F,* and *K,* are likewise 120° apart and lead to the external circuit fed by A^1, B^1, and C^1.

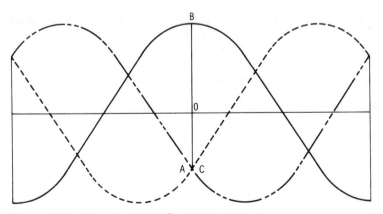

Figure 23-8 Three emfs 120° apart in phase.

This connection actually reverses winding *G-H* in relation to windings *E-F* and *I-K.* Thus the results are a wye connection as illustrated in Figure 23-5A, and sine waves as illustrated in

Figure 23-8, in which it may be observed that there is an equal current in both directions at the same time.

If winding *G-H* had not been reversed, sine waves as illustrated in Figure 23-7 would have resulted. There is no balance and the economy wouldn't be as good as that obtained in Figure 23-8. This results in a saving in copper.

Figure 23-9 graphically shows what happens when one phase of a three-phase alternator is reversed, to obtain sine waves as in Figure 23-8 instead of those shown in Figure 23-7.

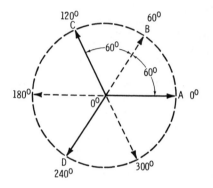

Figure 23-9 Reversal of one phase to alter the 60° relation of the phases to 120°.

Phases *A*, *B*, and *C* are 60° apart. This would give sine waves as in Figure 23-7 and an unbalanced current and voltage relationship. By taking *OB* and reversing it to *OD*, this places all phases in a 120° relationship.

Three-phase windings may be connected either wye or delta. Figure 23-10 shows the actual connections in an alternator for a delta-connected winding. It should be noted that phase-winding connections to *A* are reversed as opposed to phases *B* and *C*.

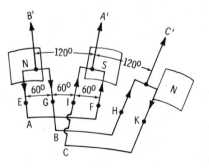

Figure 23-10 Three-phase alternator winding—connected delta.

The internal connections of an alternator must be visually checked to ascertain if the winding is connected in wye or delta.

Sometimes a wye-connected alternator may be connected four-wire, that is, the three-phase wires and the common internal connection are all brought out externally.

Questions

1. Sketch a three-wire, two-phase connection.

2. Sketch a four-wire, two-phase connection.

3. Give voltage relations for both of the above.

4. Windings of a three-phase alternator are connected so that the voltages are apart. (True or false?)

5. Which is the most economical: a single-phase or a three-phase alternator?

6. Draw a delta connection.

7. Draw a wye connection.

8. Sketch three-phase sine waves, with voltages 120° apart.

9. Sketch three-phase sine waves with voltages 60° apart.

Chapter 24

Power in Polyphase Circuits

In a three-phase machine-connected wye (Y), the emf between any two lines is equal to the voltage of one phase times the square root of 3.

In Figure 24-1, lines 1 and 2 are supplied by phases *A* and *B* of the alternator winding. As described in Chapter 14, phases *A* and *B* are 120° apart, so the voltage of the two windings can't be added by arithmetic. It may be noted from this illustration that the emf of any one phase winding is 277 volts and, as stated, the voltage of one phase times $\sqrt{3}$ is the voltage between two lines. Thus, $277 \times 1.73 = 479$ volts, line to line.

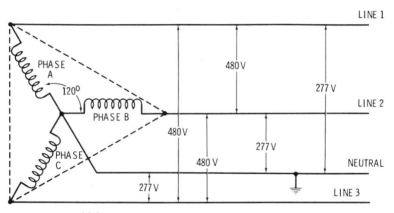

Figure 24-1   Voltages in a wye system.

This may be arrived at mathematically through the use of trigonometry. In Figure 24-2, it may be observed that the voltage, *E*, due to two phases in series 120° apart, is to the sine of the angle opposite it, which is 120°, as the voltage of one phase, *A,* is to the sine of the angle opposite it, which is 30°. The sine of 120° is the same as the sine of 60°. Thus

$$\frac{E}{A} = \frac{\sin 120°}{\sin 30°} = \frac{\sin 60°}{\sin 30°} = \frac{0.866}{0.500} = 1.732 = \sqrt{3}$$

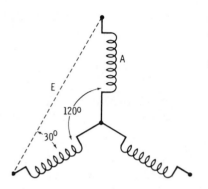

Figure 24-2   Calculating voltage delivered by two-phase windings 120° apart when connected in series.

Three-Phase Wye Connection

The relation of voltages, currents, and power in separate phases to the total power of the system in both wye and delta connections may be considered now:

For a Wye Connection at 100% PF

$$E = e\sqrt{3} \tag{1}$$

$$e = \frac{E}{\sqrt{3}} \tag{2}$$

$$i = I \tag{3}$$

$$p = eI \tag{4}$$

$$p = I = \frac{E}{\sqrt{3}} \tag{2) \& (4}$$

$$P = 3p = 3\frac{IE}{\sqrt{3}} \tag{5}$$

$$P = IE\sqrt{3} \tag{6}$$

where

p = volt-amperes per phase
P = total power in system in volt-amperes
E = voltage between any two line wires
e = emf of one phase
i = current per phase
I = current in each line wire

In analyzing these formulas, it will be seen that in a three-phase system, the total power P is *not* the current I in each line wire multiplied by each of the voltages between adjacent line wires and then totaled. Due to the angles of displacement, the total power P is the current I in a single-line wire multiplied by the voltage E between any two lines times the square root of 3, which is formula (6) above.

The power in a three-phase system is always equal to the sum of the power in the three separate phases (5). To verify, use Figure 24-3, in which the current per phase is 10 A and the voltage per phase is 1000 V; so, $p = ei = 1000 \times 10 = 10{,}000$ VA (formula (4)). Now $10{,}000 \times 3 = 30{,}000$ VA total. To check this, note that $E = e\sqrt{3} = 1000 \times 1.73 = 1732$ V (1), and $P = IE\sqrt{3}$ (6), and I is the same as i in a wye connection, so $P = 10 \times 1732 \times \sqrt{3} = 30{,}000$ VA.

If the load has 100% PF, then the watts and volt-amperes will be equal. If the PF is less than 100%, the formula for P in watts will be

$$P = \sqrt{3} \times IE \cos \phi$$

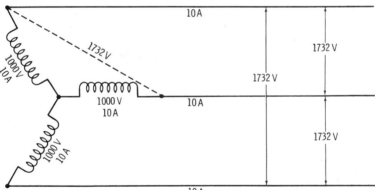

Figure 24-3 Relationships of voltage and currents delivered by a wye-connected alternator.

Lest there be some confusion, apparent power is rated in VA or kVA and true power is rated in watts (W) or kilowatts (kW). A thorough understanding of kVA and kW is essential to a thorough understanding of power in AC circuits.

Three-Phase Delta Connection

Current and voltage relationships in a delta-connected (Δ) alternator or other AC delta connections are different than in a wye-connected alternator or other wye-connected circuits.

$$I = i\sqrt{3} \tag{1}$$

$$i = \frac{I}{\sqrt{3}} \tag{2}$$

$$e = E \tag{3}$$

$$p = ei \tag{4}$$

$$p = Ei \tag{5}$$

$$P = 3p = 3\frac{IE}{\sqrt{3}} \quad P = IE\sqrt{3} \tag{6}$$

where the letter symbols have the same meaning as before.

The formula for total power in VA in a delta connection is the same as for a wye connection. The square root of 3 is a factor used in both wye and delta connections. With wye connections, $\sqrt{3}$ is a factor of the line voltage, and with delta connections the $\sqrt{3}$ is a factor with line current. Thus for Figure 24-4,

$$I = i\sqrt{3} = 10 \times 1.732 - 17.32 \text{ amperes} \tag{1}$$

$$E = e \quad \text{or} \quad 1000 \text{ volts} = 1000 \text{ volts} \tag{3}$$

$$P = IE\sqrt{3} = 17.32 \times 1000 \times 1.732 = 30,000 \text{ VA} \tag{6}$$

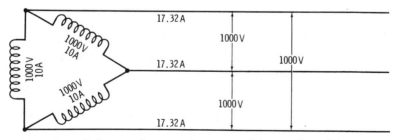

Figure 24-4 Relative voltages and currents delivered by a delta-connected alternator.

If the PF is less than 100%, then

$$P \text{ in watts } = IE \cos \phi \sqrt{3}$$

Note that P is equal to the total power in the system and is in VA, or apparent power. To get kilovolt-amperes, kVA = VA/1000. At 100% PF, P would be both volt-amperes and watts, but with other than 100% PF, P in VA has to be multiplied by cos ϕ to get watts.

Figure 24-5 will be used to compare power loss in a three-phase circuit as compared to a one-phase two-wire circuit. The resistance (Ω) of each line times I^2 (10^2) equals 200 VA lost in each line, or $200 \times 3 = 600$ VA lost in all three lines.

$$\text{Total } P = IE\sqrt{3} = 10 \times 1000 \times 1.732 = 17,320 \text{ VA}$$

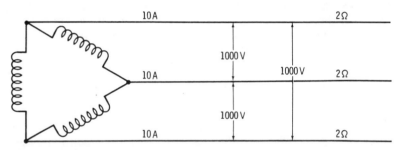

Figure 24-5   Power lost in three-phase transmission.

To carry this same power in a two-wire, one-phase line, $I = P/E = 17,320/1000 = 17.32$ A per line. The total loss in the three-phase system was 600 VA, so for the same power in the one-phase system, $600/2 = 300$ VA lost per conductor. Now $R = P/I^2$ so $R = 300/17.32^2 = 1$ ohm. This means that the one-phase lines would have twice the cross-section of the three-phase lines, so each conductor will weigh twice as much for the one-phase line as for the three-phase lines. See Figure 24-6.

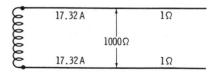

Figure 24-6   Copper required for a one-phase system.

Economy and Phases

The following are comparisons of weights of copper required for single-phase and three-phase power systems for carrying equal amounts of power the same distances and with the same losses. Two-wire, one-phase will be used as the base for comparison. The amperes and watts or volt-amperes won't be called out, as they are all equal. The base voltage will be 100 volts. See Figure 24-7.

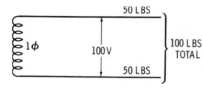

Figure 24-7 Total copper required for a one-phase transmission.

It was just shown that the conductors of a three-phase system were half the cross-section of those for a one-phase system for the same power delivered; Figure 24-8 illustrates the comparison of a three-phase circuit to the one-phase circuit.

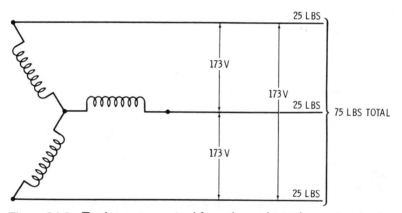

Figure 24-8 Total copper required for a three-phase, three-wire circuit.

The following will illustrate the economy of a three-phase, four-wire system. The economy of copper in a transmission line varies as the square of the applied voltage, as we saw in a previous chapter. In Figure 24-9, there are still 100 volts between any phase and the neutral and 173 volts between phases. So $173^2 = 29,929$ and $100^2 = 10,000$, or a 3-to-1 ratio. So the conductors will weigh as shown in Figure 24-9.

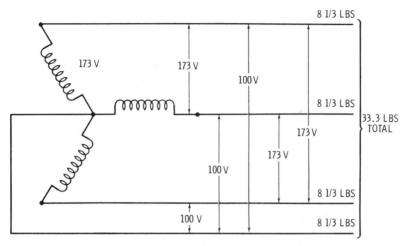

Figure 24-9   Total copper required for a three-phase, four-wire system.

Neutral Current in Three-Phase, Four-Wire Systems

On three-phase, four-wire systems used for discharge lighting, the neutral current with two phases will be considered as being the same value as the phase current. This is due to the third harmonic of the line frequency. (Harmonics are multiples of the fundamental frequency.) This is illustrated in Figure 24-10A, where only the second and third harmonics are shown.

Figure 24-10B shows the third harmonic with the three fundamental phases. By studying this illustration, it is plain that the third harmonic of each phase adds to the third harmonic of the other phases. This is the reason that the *NEC* won't permit the derating of the neutral of a three-phase, four-wire circuit when discharge lighting is supplied.

For three-phase, four-wire circuits with other than discharge lighting, the following vector methods of calculation of the neutral current will be found to be useful.

Problem

Using Figure 24-11, draw vectors of the currents as in Figure 24-12, scaling the lines to represent the amperes per phase, and the angles of phase displacement: $OA = 10$ amperes in phase A, $OB = 10$ amperes in phase B, and $OC = 10$ amperes in phase C. Draw the line ON, which comes out 10 amperes and diametrically opposite

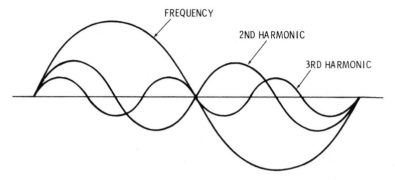

FREQUENCY

2ND HARMONIC

3RD HARMONIC

(A) Fundamental, second, and third harmonics.

(B) Third harmonic is in phase with the third harmonics of the other phases.

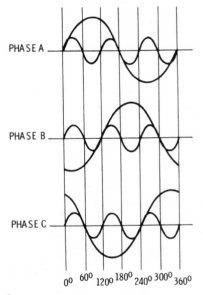

PHASE A

PHASE B

PHASE C

0^0 60^0 120^0 180^0 240^0 300^0 360^0

Figure 24-10 Harmonics of the phase current.

to line *OC*. This is 10 amperes in the neutral for phases *A* and *B*, but *OC* is diametrically opposite to *ON*, so *OC* and *ON* will cancel and the resultant neutral current for a balanced system will be zero.

For a three-phase, four-wire system with two phases balanced and one phase current variable, as in Figure 24-13, draw a vector as in Figure 24-14 except *OC* will be marked in various current

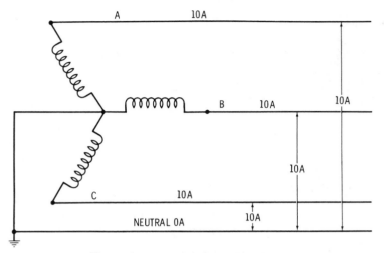

Figure 24-11   Three phases with balanced loads.

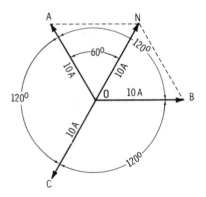

Figure 24-12   Vectors of currents in Figure 24-11.

lengths in proportion to the currents for OC as shown in Figure 24-14. For example, suppose OC takes the values 3, 6, and 10 amperes.

$OC_2 = 3$ amperes   so neutral current = 7 amperes
$OC_1 = 6$ amperes   so neutral current = 4 amperes
$OC = 10$ amperes   so neutral current = 0 amperes

As before, the portion of OC diametrically opposite to ON is subtracted from ON, giving the neutral current for any unbalances.

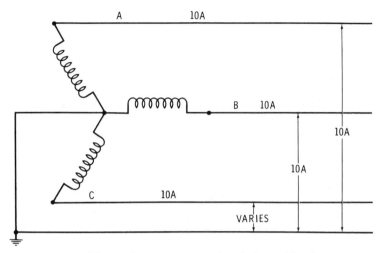

Figure 24-13 Three-phase system with unbalanced loads.

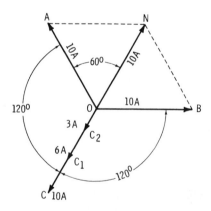

Figure 24-14 Vector diagram of currents of the circuit in Figure 24-13.

The same method will be used for unbalanced loads of all phases (see Figure 24-15). Draw a parallelogram $AOCN$ with line NC parallel to line AO, and NO will be measured to find the current in ON. In this case, ON is 5.2 amperes.

All of this may be worked out by simple trigonometry. From the law of cosines,

$$ON^2 = OC^2 + OA^2 - 2 \times OA \times OC \times \cos NCO$$

To find angle NCO, observe that angles AOC and CNA are equal and 120° each. Since the interior angles of a quadrilateral

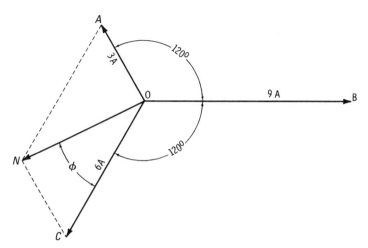

Figure 24-15   Finding current in three-phase unbalanced load.

total 360°, the sum of angles NAO and NCO must be 360° − 2(120°) = 120°. But angles NAO and NCO are equal; therefore, angle NCO is 120°/2=60°. Thus

$$ON^2 = 6^2 + 3^2 - 2 \times 6 \times 3 \cos 60°$$

$$= 45 - 36 \times 0.5 = 45 - 18 = 27$$

So

$$ON = \sqrt{27} = 5.2 \ A$$

In similar vector fashion, ON may be added to OB.

The same method may be used for the other phase currents which were just covered.

Formulas

$$V/A = \sin 120°/\sin 30° = \sin 60°/\sin 30° = 1.732 = \sqrt{3}$$

For Wye Connections

$$E = e\sqrt{3}$$

$$i = I$$

$$e = E/\sqrt{3}$$

$$p = eI$$

$$p = JE/\sqrt{3}$$

$$p = 3p = 3IE/\sqrt{3} = IE\sqrt{3}$$

For less than 100% PF,

$$P = \sqrt{3} \times IE \times \cos \phi$$

For Delta Connections

$$I = i\sqrt{3}$$

$$i = I\sqrt{3}$$

$$e = E$$

$$p = ei$$

$$P = Ei$$

$$P = 3p = \frac{3IE}{\sqrt{3}} = IE\sqrt{3}$$

For less than 100% PF,

$$P = \sqrt{3} \times IE \cos \phi$$

$$R = P/I^2$$

Law of Cosines:

$$a^2 = b^2 + c^2 - 2bc \cos A$$

where sides *a*, *b*, and *c* of triangle *abc* are opposite angles *A*, *B*, and *C*, respectively.

Questions

1. In a wye system, if the line voltage is known, give the formula for the phase voltage.
2. In a wye system, give the formula for the line current.

3. In a delta system, give the formula for the line current.

4. In a wye system, give the formula for the phase voltage.

5. In a delta system, give the formula for the phase voltage.

6. Give the formula for total power in VA for three-phase circuits.

7. Give the formula for total power in watts for three-phase circuits.

8. Knowing the VA, how do you find the kVA?

9. Knowing the watts, how do you find the kilowatts?

Chapter 25

Transformer Principles

Chapter 16 covered electromagnetic induction. In transformers, the theory of mutual induction is the basis of their operation.

Induction Coil

Before discussing transformers as they are commonly thought of, the induction coil may be covered. It has the same fundamental principle as the transformer, except the input is usually DC, which is interrupted by a vibrator or some other means.

A bundle of soft iron wires usually forms the core, as in Figure 25-1. The primary winding is composed of relatively few turns of wire, and the secondary conductor is wound outside over the primary conductors and consists of many turns of a smaller-size wire than the primary.

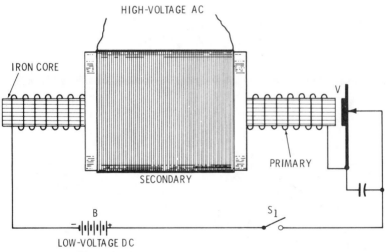

Figure 25-1 Typical induction coil.

Battery *B* supplies the electricity to the induction coil. Since DC causes induction for the initial charge only, a means must be supplied for interrupting the supply from the battery very quickly. This may be accomplished by a magnetic vibrator and points of *V*. When switch S_1 is closed, the primary is energized, and the armature of vibrator V is pulled to the core, opening the points of V, deenergizing the primary; the capacitor C absorbs the current that would cause sparking at the points, causing a more rapid collapse of the magnetic field and thus a higher induced emf in the secondary.

Figure 25-2 illustrates the action that takes place. When the primary is energized, the primary current rises from A to B. This in turn induces a small emf in the secondary in the opposite direction, as shown by AFC. As the time AC for the rising of current in AB is considerable, the magnitude of the induced emf, GF, in the secondary is small. When the current rising against the emf of self-induction reaches its maximum at B, and is momentarily stationary, the induced emf falls to C. The interrupter opens and breaks the circuit, and the current collapses from B to E, and with it the magnetic flux. The capacitor ensures a very short time interval from C to E, during which the flux collapses. The magnitude of the induced emf in a negative direction will soar to a great height as from C to P, and when the current finally reaches zero at point E, the secondary emf has collapsed from P to zero.

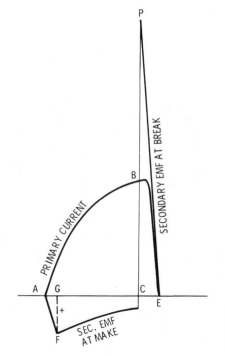

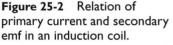

Figure 25-2 Relation of primary current and secondary emf in an induction coil.

In an induction coil, it is not necessary to apply an AC emf, as the vibrator or interrupter causes a pulsating DC. The secondary emf, *AFC*, is of little value as it is so feeble.

There are many variations of the induction coil. The intent here was merely to show the principle. The induction coil with which everyone is most familiar is probably the ignition coil on the automobile. (It would be good to observe that there is no such item as a DC transformer.)

Joseph Henry devised the first transformer in 1832. The first practical transformer was patented in England in 1882 by Gaulard and Gibbs. The American rights were purchased by Westinghouse in 1886. The first transformer to be built in this country was built by William Stanley in 1885, an employee of the Westinghouse Company.

Transformer Types

The efficiency of induction coils is very low, but the efficiency of transformers is in the range of 92 to 99%. Large transformers often are nearly 99% efficient. Beyond a doubt, the transformer is one of the most efficient pieces of electrical equipment ever made.

Transformers are for transforming AC voltages from one voltage to another. This voltage transformation may be either up or down in value. Both step-up and step-down transformers are alike: One is for stepping up the voltage, and the other is for stepping down the voltage. Figure 25-3 illustrates both a step-up transformer and a step-down transformer, and they may be identical types of transformers.

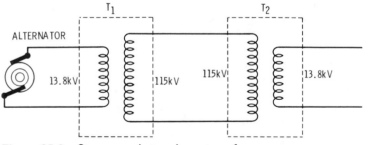

Figure 25-3   Step-up and step-down transformers.

In Figure 25-3, 13.8 kV (kilovolts) are delivered to transformer T_1, which steps up the voltage to 115 kV for transmission to a distant point, transformer T_2 (which could be an identical transformer to T_1), where the voltage is transformed down from 115 kV to 13.8 kV for distribution.

Principles of Operation

Transformers are sometimes termed *static transformers* since they have no moving parts. The principle of a transformer is illustrated in Figure 25-4.

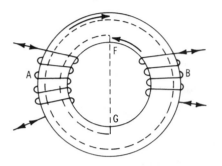

Figure 25-4 Principle of a transformer.

Assume that 100 volts AC are impressed on winding *A*, wound around a soft iron ring *C*, and the coil contains 300 turns. A magnetic flux is set up by the current in *A* within the soft iron core *C*. This flux is set up around each turn or convolution of coil *A*. The flux lines are like rubber bands that expand outward, cutting each turn of winding *A*, expanding until they cut each turn of coil *B*, before they eventually occupy the cross-section of core *C*.

As the impressed voltage and the current in coil *A* rise and fall, the flux in turn rises and falls, cutting the turns of coil *B*. From the principle of induction, the rising and falling of the flux cutting coil *B* induces an emf in coil *B*. There is also a counter emf induced in coil *A*. This emf and the emf induced in coil *B* are in opposition to the impressed emf in coil *A*.

Now, if coil *B* is open, that is, has no load connected, there will be no flux generated by coil *B* in opposition to the flux generated by the impressed emf and current in coil *A*. The counter emf in coil *A*, however, in opposition to the impressed emf on coil *A*, opposes the changes in current and tends to keep the flux oscillating with the frequency of the impressed emf.

This may be compared to the shunt motor as covered in Chapter 19, where the effect of counter emf was illustrated in Figure 19-5 with the accompanying explanations. Namely, the rotating armature generated a counter emf almost equal to the impressed emf, so that the impressed emf delivers just enough current to keep the armature going.

The difference here is that the transformer windings are stationary, while the flux is oscillating.

Windings and Voltages

The winding that receives the impressed emf is always termed the *primary* winding, regardless of the impressed voltage; and the winding that receives the induced emf, in this case winding *B*, is the *secondary* winding. There is very often confusion in that the high-voltage winding is called the primary and the low-voltage winding is called the secondary. As may be seen, this is not the case. The input side is the primary, and the output side is the secondary.

The voltages of a transformer are proportional to the number of turns in the windings. Thus, if in Figure 25-4 winding *A* has 300 turns and an impressed voltage of 100 volts, and winding *B* has 150 turns, then winding *B* will have an induced voltage of 50 volts.

Each convolution or turn of transformer coils is cut by magnetic flux four times per cycle: first, when the flux is rising; second, when the flux is falling to zero; third, when the flux is rising in the reverse direction; and fourth, when the flux is falling to zero.

The average voltage that is induced in each winding with a flux having a maximum value of ϕ will be

$$E_{avg} = 4\phi nT/10^8$$

where

E_{avg} = average emf induced in each winding
ϕ = maximum magnetic flux
n = frequency in hertz
T = number of turns in coil
4 = constant: number of times each turn is cut by flux per cycle
10^8 = constant to reduce absolute lines of force for conversion of terms to practical volts

The rms voltage or effective voltage, as read on the voltmeter, equals 1.11 times the average voltage. So, for the effective voltage instead of the average voltage, the 4 in the above formula becomes 4 × 1.11, or 4.44. Thus,

$$E_{rms} = 4.4\phi nT/10^8$$

It was shown that for the transformer in Figure 25-4, with no load on winding B, the current into winding A was limited to a low value by the counter emf induced in that winding.

If, however, AC voltage is impressed across the primary of an iron-core transformer, a current surge may occur. The reason for this is as follows. The voltage to be impressed has associated with it a flux curve, which is simply the flux that would occur were the AC voltage impressed on the primary. The flux curve, in its relation to its steady-state axis, is determined by two factors: (1) There is no current at the instant of application of voltage, thus the instantaneous value of the flux is zero; and (2) the rate of change of flux must be that required to cause the desired voltage to be induced (approximately equal to the applied voltage). Thus, the flux may reach a peak density of as much as twice the steady-state value. The result of this is complete core saturation, i.e., a condition in which a great change in current is required to produce even a small change in flux. The current necessary to produce the required flux may therefore be very great.

The inrush of current exists only for the first half-cycle and then drops off rapidly during subsequent cycles, thus reaching the normal charging current in two to three cycles. The maximum peak of inrushing current can reach 30 to 100 times the normal line current. As an example, a transformer with 100 volts impressed and 0.01-ohm primary resistance could reach 10,000 amperes of inrush current.

If the switch is closed and the voltage wave value is maximum, the inrush current would be very low. Thus, the inrush current is dependent upon the point of the voltage wave where the switch is closed.

With three phases, the voltage waves are 120° apart so that there will always be an inrush of current upon closing the switch. Therefore, the inrush current must be considered when determining the primary overcurrent protection.

If a load is added to winding B of Figure 25-4, this load will draw current from winding B. Thus, the current in B, the flux of which opposes the flux of the primary current, reduces the counter emf in A, which allows winding A to take more current from the source of supply in direct proportion to the secondary current. The reverse action takes place if the current load on winding B is removed or reduced.

From this, it may be observed that the primary winding of a transformer draws current in proportion to the current load on the secondary.

A transformer was compared in some respects to a shunt motor; however, the efficiency and regulation of a transformer are much superior to those of a shunt motor.

Earlier it was stated that the primary voltage and turns were proportional to the secondary voltage and turns. Thus, there would be one winding of 100 volts and 300 turns, as opposed to 50 volts and 150 turns on the other winding. This is fundamentally true of an ideal transformer with no losses, but, practically, windings must be added or subtracted from one coil or the other due to losses in flux and the like.

So far the current relationships between windings were merely mentioned as being proportional. They are inversely proportional. Take a transformer with a 100-volt primary and 5 amperes full load; this would draw 500 VA. The secondary has 50 volts and, neglecting efficiencies, the secondary would be carrying 500 VA also. So, $I = P/E = 500/50 = 10$ amperes.

Recapping the analysis of the ratio of turns, voltage, current, and power is the illustration in Figure 25-5. Here the high side has 100 V and 300 turns. So, 100 V/300 T = ⅓ volt per turn. The low side has 50 V and 150 turns. So 50 V/150 T = ⅓ volt per turn. Thus, the voltage-and-turns ratios are equal.

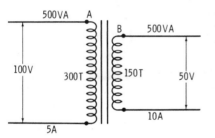

Figure 25-5 Ratio of turns, voltages, currents, and power in transformers.

Winding A carries 500 VA and so does winding B, from which it may be observed that primary and secondary windings carry the same power.

As to the current in the windings, for winding A

$I = P/E = 500\ VA/100\ V = 5$ amperes

Winding B has

$I = P/E = 500\ VA/50\ V = 10$ amperes

From this it is found that the ratio of primary turns to secondary turns is in inverse proportion to the ratio of primary current to secondary current.

Formulas

$$E_{avg} = 4\phi nT/10^8$$

$$E_{rms} = 4.4\phi nT/10^8$$

Questions

1. Induction in transformers is induction. (True or false?)
2. Sketch an induction coil and explain its operation.
3. Is there such an item as a DC transformer?
4. A transformer transforms power from one quantity to another. (True or false?)
5. What is the purpose of transformers?
6. Describe the difference between a step-down and a step-up transformer.
7. The primary of a transformer is always the high side. (True of false?)
8. Do transformers have moving parts?
9. Explain the principle of transformer action.
10. Explain the principle of transformer regulation.
11. What is inrush current?
12. What is the normal inrush current range?
13. The current ratio of the two windings of a transformer is proportional to the ratio of_____.
14. The voltage ratio of a transformer varies as the_____.

Chapter 26

Transformer Facts

Transformer cores are made up of thin strips of a very high grade of electrical steel. Usually it takes 40 to 50 such strips to make up one inch in thickness. These strips are called *laminations*. Each lamination has an oxide coating on one side for insulation, or one side may be coated with a varnish. The purpose of laminations, which is to reduce eddy currents, was covered in Chapter 16.

The losses in a transformer cause heating. In Chapter 25 it was explained that transformers have very high efficiencies, but there are losses that have to be considered:

1. Iron losses, namely, eddy currents and hysteresis.
2. Copper losses, which are I^2R and I^2Z losses.
3. There are also magnetic flux losses, due to leakage of the flux and the magnetic reluctance of the steel core.

Transformer losses must be considered, as they enter into the cost of electricity. Large industrial users are usually on what is termed a primary rate. This means that they may own the transformer, and the electricity is metered at the primary side of the transformer.

When the transformers have a voltage that is too high to meter, then the power will be metered on the secondary side and the transformer losses will be calculated and added to the power consumption.

Sometimes the transformer losses are figured in the rate structure.

The most common configurations for transformer cores are E and I, U and I, and I laminations. In cutting or stamping out the laminations, these three types make 100% use of all the steel sheets from which the laminations are being cut.

The laminations for small and medium-size transformers are usually made up of E and I laminations. See Figures 26-1 and 26-2. Figure 26-1A shows the final stamping of E and I laminations. Figure 26-1B shows proportions (these are not meant to be dimensioned) of the laminations. Also shown are staggering of laminations during stacking (Figure 26-1C), with the windings and the flux paths, while Figure 26-2 shows the layout of the stampings. Note that there is no steel lost from the stampings.

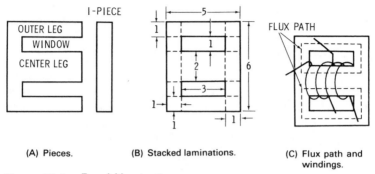

(A) Pieces.   (B) Stacked laminations.   (C) Flux path and windings.

Figure 26-1 *E* and *I* laminations.

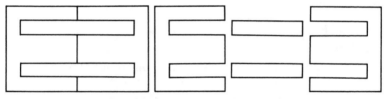

Figure 26-2 How *E* and *I* laminations are laid out for stamping.

The design must be such that the core is not saturated with magnetic flux. Operation at too high a voltage will cause flux saturation and this will affect the transformer operation. Note that in the *E* and *I* laminations, the core is twice the dimensions of the outer legs.

Figure 26-3A shows the shape of the *U* and *I* laminations. Figure 26-3B shows the staggering of the laminations on assembly, and Figure 26-3C shows the flux path.

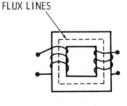

(A) Laminated pieces.   (B) Laminations assembled.   (C) Windings and flux path of core.

Figure 26-3 *U* and *I* laminations.

Figure 26-4A shows the *I* laminations. Figure 26-4B shows their staggered joints while being assembled, and Figure 26-4C shows the flux path of the *I*-laminated core.

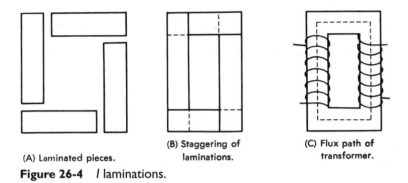

(A) Laminated pieces.

(B) Staggering of laminations.

(C) Flux path of transformer.

Figure 26-4   *I* laminations.

There are other core configurations, one of which is the circle or doughnut core.

During the core assembly, it is essential that care be taken in assembly and the joints be as tight as possible, wedged and secured or clamped to prevent noise from vibrating laminations, to keep the sound levels of the transformer low and to keep the path for the flux at as low a magnetic reluctance as possible. Burr-free laminations also aid in keeping sound levels low.

During each half-cycle the laminations tend to change their dimensions. This is called *magnetostriction* and is due to the molecules of the steel acting as very small magnets and changing their positions.

After the coils are assembled on the core, the core and windings are dipped into insulating varnish and baked. Before dipping they must be thoroughly dried.

There are many facts concerning transformers that are common to all transformers. These will be covered here not in any particular order of importance, as they are all important.

Transformer Impedance

Impedance in a transformer is composed of both resistance and the inductive reactance of the winding: $Z = \sqrt{(R^2 + X_L^2)}$.

The voltage drop in a transformer from no load to full load is due mostly to the resistance of the windings (see Figure 26-5), as the resistive component of the winding is in phase with the

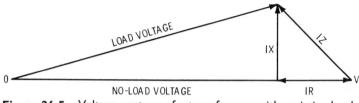

Figure 26-5 Voltage vectors of a transformer with resistive load.

voltage, and the reactive component of the voltage (IX) is 90° out of phase with the impressed voltage. The vectors in Figure 26-5 may be used to illustrate this. IR subtracts directly, as it is in line with the no-load voltage, while IX subtracts at a 90° angle. The resultant voltage drop will be IZ. OV represents the no-load voltage and

$$IZ = \sqrt{(IR)^2 + (IX)^2}$$

Transformer nameplates have the transformer impedance indicated.

Short-circuit current is the available current that a given transformer will pass on short circuit of the secondary. This is dependent upon the transformer's impedance and may best be illustrated by examples.

100-kVA, 480-volt, one-phase transformer has an impedance of 5%. How many amperes of fault current will be available upon short circuit?

$$\frac{100 \text{ kVA} \times 1000}{480 \text{ V}} = 208.3 \text{ amperes (full load)}$$

$$I \text{ (short-circuit)} = \frac{100\%}{\text{Percent Impedance}} \times I \text{ (full load)}$$

$$I_{SC} = \frac{100\%}{5\%} \times 208.3 = 20 \times 208.3 = 4166 \text{ amperes}$$

If the transformer above were a three-phase transformer, then

$$\frac{100 \text{ kVA} \times 1000}{480 \text{ V} \times 1.73} = 120.4 \text{ amperes}$$

$$I_{SC} = \frac{100\%}{5\%} \times 120.4 = 20 \times 120.4 = 2408 \text{ amperes}$$

Transformers in Parallel

Transformers may be paralleled if more capacity is required over that which one transformer will supply. There are very definite guides that must be followed:

1. The combined kVA rating of the transformers shall be large enough to handle the load.
2. The transformers must be connected so that their polarities are the same. (Polarities will be covered a little later.)
3. The transformers shall have identical voltage and frequency ratings.
4. The impedances and percentages of resistance and reactance shall be identical.
5. They shall have the same identical tap ratings.
6. With three-phase banks, the phase rotations shall be identical, such as phase A', B', and C to phases A', B', and C, and not phases A', B', and C to phases A', C, and B, etc. More will be covered on phase rotation in a future chapter.
7. The phase banks shall have the same phase displacement angles, such as delta-delta, wye-wye, delta-wye, or wye-delta. If a delta-delta is paralleled with a delta-wye, there will be a 30° phase displacement. This will be explained in a later chapter.

Transformer Insulation

Most types of insulation must be impregnated with an insulating varnish, which

1. Improves resistance to moisture
2. Improves temperature ratings
3. Improves dielectric strength
4. Fills in small voids
5. Gives mechanical strength
6. Molds the assembly into one solid mass to prevent movement
7. Also aids in keeping down the noise from lamination vibrations

There are a number of insulation classes. The insulation class picked should take into consideration the application and location of the transformer that is to be used. The inside of the winding usually gets the hottest and there is about 10° Celsius (centigrade) difference between this hot-spot temperature and the average temperature. A 10% rise in temperature cuts the insulation life in half.

NEMA and ANSI insulation classes are

> 105°C rise—Class A
> 130°C rise—Class B
> 155°C rise—Class F
> 180°C rise—Class H

Types of Transformers

There are three basic types of transformers:

1. *Dry Type:* This is the type that is not immersed in a liquid dielectric.

2. *Oil-filled:* This has the core and windings immersed in a high grade of insulating mineral oil, which serves as a coolant and dielectric. Carbons and water are heavier than this oil, so they settle to the bottom. When oil is drawn off for checking its dielectric strength, the test sample is taken from the bottom of the tank.

3. *Askarel-filled:* Askarel is no longer manufactured because of its toxicity. (Askarel contained PCBs.) Nonetheless, you may come across an older transformer filled with askarel, which acts as a coolant and dielectric as does oil. Water and carbons are lighter than askarel, and thus rise to the top. Therefore, test samples are taken from the top of the transformer.

Quite often transformers are tightly sealed and nitrogen gas is added under a low pressure to keep outside air and moisture from entering the case, which prevents contamination.

Effects of Altitudes

Transformers are designed and rated for altitudes from sea level to 3300 feet. As altitudes increase, the atmosphere becomes less dense, resulting in less effective cooling. Therefore, transformer rated capacity must be derated 0.3% for each 330 feet over 3300 feet of altitude. This is 1.0% derating for each 1000 feet over 3300 feet.

Ambient Temperature

Ambient temperature is the temperature in the location where the transformer will be located. High ambient temperatures will raise the insulation temperatures and shorten the life of a transformer.

The kVA rating of a transformer should be derated 0.4% for each 1°C over 30°C average ambient temperature for 150°C insulation.

Transformer Polarity

Transformers are often marked either "Additive Polarity" or "Subtractive Polarity." The leads on the high side are usually marked H_1, H_2, H_3, etc., and the leads on the low side are marked with an X, such as X_0, X_1, X_2, X_3, etc. This is done for phasing out, paralleling, and connecting transformers for different voltages. Sometimes the markings may have been changed or marked incorrectly, so it may become necessary to check the polarity.

Figure 26-6 gives the illustration of additive polarity. Figure 26-6A shows the actual direction of the windings. Note that both the high and low sides are wound in the same direction, and in Figure 26-6B the arrows show the direction of the impressed voltage on the high side and direction of the induced voltage on the low side.

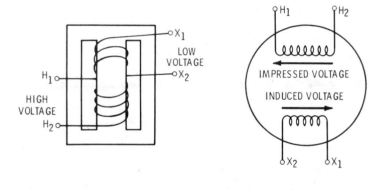

(A) Direction of windings.   (B) Direction of voltages.

Figure 26-6   Additive polarity transformers.

Figure 26-7 illustrates the method of checking polarity. Connect one side of one winding to one side of the other winding as shown and mark the high-side connection H_1. Connect a voltmeter between the other low-side lead and the other high-side lead. Apply the primary voltage as shown, and if the polarity is additive, the voltmeter

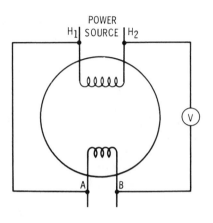

POWER
H₁| SOURCE |H₂

A B

Figure 26-7 Checking transformer polarity.

Figure 26-7 Checking transformer polarity.

V will read the primary voltage plus the secondary voltage. If this shows additive, mark the leads as shown in Figure 26-6B. A study of the direction arrows in Figure 26-6B and Figure 26-7 will show why the voltages add together.

If the voltmeter reads less than the primary voltage, the secondary leads should be marked as shown in Figure 26-8 and the transformer would be of subtractive polarity. A study of Figure 26-8 will show why the voltmeter reads less than the voltage of the power source.

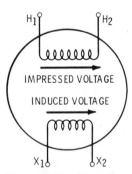

H₁   H₂

IMPRESSED VOLTAGE

INDUCED VOLTAGE

X₁   X₂

Figure 26-8 Transformer with subtractive polarity.

Termination Marking

Figure 26-9 illustrates a dual-voltage one-phase transformer. The high side for this coverage won't be dual voltage, but the low side will be dual voltage. Figure 26-9A shows the basic idea, and, as noted, the X_2 and X_3 leads are crossed internally in the transformer, as shown in Figure 26-9B.

Figure 26-10A illustrates the low-voltage side connected with the windings in series. The crossing of X_2 and X_3 internally has no effect on this 120/240-volt connection, but in Figure 26-10B, the crossing of the low side internally does affect the external connections for 120 volts only. It is not necessary to cross the leads externally, which aids in avoiding mistakes in the field.

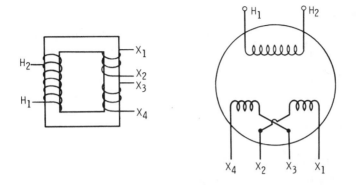

(A) Basic construction.   (B) Leads X_2 and X_3 crossed internally.

Figure 26-9   Single-phase transformers with dual low-voltage windings.

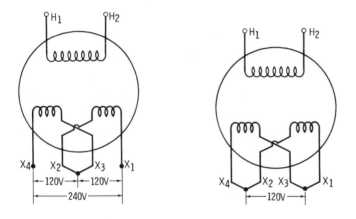

(A) Windings in series.   (B) Windings in parallel.

Figure 26-10   Connections of low-voltage windings.

Formulas

$$Z = \sqrt{R^2 + X^2}$$

$$IZ = \sqrt{(IR)^2 + (IX)^2}$$

$$I \text{ (short circuit)} = \frac{100\%}{\text{Percent Impedance}}$$

Questions

1. Transformer cores are made of cast iron. (True or false?)
2. Why are laminations insulated from each other?
3. What losses occur in transformer cores?
4. Show by sketches three types of core formations.
5. How are core noises kept low?
6. Why are transformer cores and coils dipped in insulating varnish?
7. What makes up transformer impedance?
8. Give the formulas for short-circuit current.
9. In paralleling transformers, give the requirements that must be met.
10. Name the classes of transformers.
11. Give the ANSI insulation class temperatures.
12. Explain the effect of altitude on transformer output.
13. What is ambient temperature?
14. What markings appear on transformer leads?
15. Draw sketches of how to check polarity and explain.

Chapter 27

Transforming Polyphase Power

Two-phase power transforming requires two transformers. A schematic diagram of connections is shown in Figure 27-1. Either the high- or low-voltage side or both may be connected three-wire or four-wire as desired.

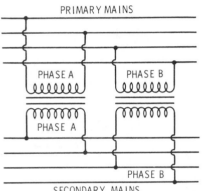

PRIMARY MAINS

PHASE A   PHASE B

PHASE A

PHASE B

SECONDARY MAINS

Figure 27-1 Two transformers connected two-phase four-wire.

Single-Phase for Three-Phase

Single-phase transformers or a single three-phase transformer may be used with three-phase power.

Three single-phase transformers are often used by utility companies on their distribution lines, whereas three-phase transformers are often used at substations.

Figure 27-2 shows three single-phase transformers in a delta-delta bank. Note that each transformer is connected as a single transformer, such as transformer No. 1 being connected independently of transformers No. 2 and No. 3. Lines *A* and *B* supply transformer No. 1 and the output is connected to *X* and *Y*, as shown in both the schematic and the vector diagram. Note in Figure 27-2 that no neutral connection is shown.

Two single-phase transformers may be connected in open-delta or V connection as shown in Figure 27-3. This gives very satisfactory

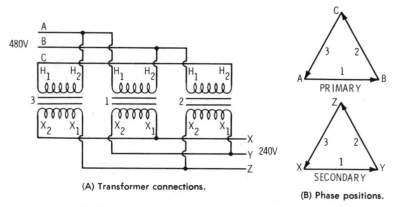

(A) Transformer connections.

(B) Phase positions.

Figure 27-2 Delta-delta connections and vectors.

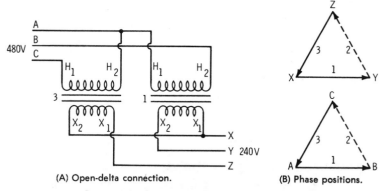

(A) Open-delta connection.

(B) Phase positions.

Figure 27-3 Open-delta connections using two single-phase transformers.

voltage transformation. Phase Y-Z is often termed the *phantom phase* and is usually a few volts higher than the other two phases. This is widely used on distribution lines. Again, as in Figure 27-2, there is no neutral shown at this point.

Two transformers in open-delta won't deliver 100% of their kVA ratings to a three-phase load, but will deliver only 86.6% of their total kVA ratings; and the two transformers can deliver only 57.7% of the load that could be carried if the third transformer were present.

One great advantage of three single-phase transformers in a three-phase bank is that under emergency conditions, two of the three transformers may be used to supply a part of the load.

Very often a neutral is used with delta-delta connections or open-delta connections. For instance, both three-phase 240 volts and single-phase 120/240 volts may be obtained from the same transformer bank. One transformer is often larger than the other two to serve the single-phase load portion, and the other transformer or transformers are smaller, to serve only their portion of the three-phase load with the larger transformer.

Figure 27-4 illustrates a four-wire open-delta connection and the resultant voltages. Note that Y to N gives 208 volts, which is 120 volts × 1.73, and results from phase relationships or angular displacements. Leg Y is often called the *wild leg,* and the *NEC* requires that be identified wherever it appears with the neutral.

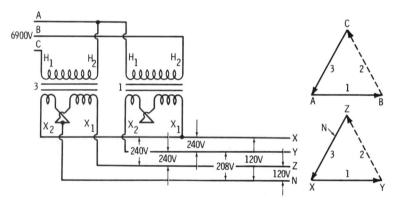

Figure 27-4   Open-delta four-wire connection and vectors.

Also, one will sometimes find one phase of a closed-delta or open-delta transformer connection grounded. This won't change any voltage from those shown in Figures 27-2 and 27-3. There is no *NEC* violation; this is merely one method of grounding the secondary of a delta bank. This method of grounding is not found too often, but it is used at times.

Three-Phase Transformers

So far this discussion has dealt with single-phase transformers connected for three-phase service. There are three-phase delta-delta

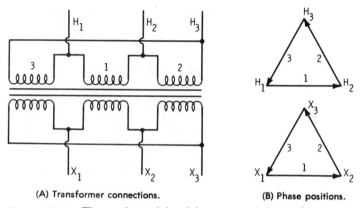

(A) Transformer connections.   (B) Phase positions.

Figure 27-5   Three-phase delta-delta connected transformer.

transformers in one enclosure. Figure 27-5 shows such a transformer, and the connections and the vectors are shown.

Delta-delta connections are very satisfactory. There are no problems with third harmonics, and there is no angular phase shift between primary and secondary.

Wye-wye connected transformers are not so popular. See Figure 27-6. This is because they have an inherent instability of the neutral. The three-phase, three-legged core types are a little more stable than the five-legged core types or three single-phase units. If the H_0 and X_0 terminals are grounded together and the load unbalance doesn't exceed 10%, they give fairly satisfactory results, but with unbalance of more than 10%, they are subject to excessive heating.

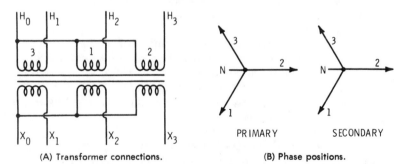

(A) Transformer connections.   (B) Phase positions.

Figure 27-6   Three-phase wye-wye connected transformer.

The most popular three-phase transformers are delta-wye. These have no third harmonics or other problems. Figure 27-7 shows a delta input and a wye output with three single-phase transformers. This will work with inputs such as 480 volts, with the output connected for 120 volts on each transformer, thus giving 208 volts three-phase and 120 volts single-phase.

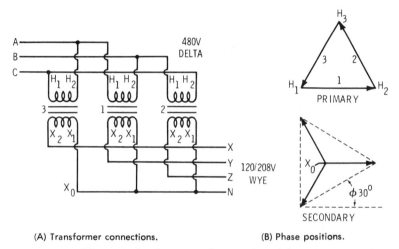

(A) Transformer connections.   (B) Phase positions.

Figure 27-7   Three single-phase transformers in a delta-wye connection.

To obtain full output, each transformer winding must carry full current at the rated voltage. Figure 27-7A shows such a connection. To clarify, Figure 27-7B is shown. If a single three-phase transformer is to be replaced by three single-phase transformers and the connection is delta-wye, the phase relationships of both current and voltage must be taken into account, due to the 120° displacements. Note from the vectors shown in Figure 27-7B that there is a 30° angular displacement between the windings, which was mentioned in the last chapter under paralleling of transformers.

Figure 27-8 shows a three-phase delta-wye transformer connection and the phase relationships.

Two-Phase to Three-Phase Conversion

The Scott connection for changing two-phase to three-phase, while not much used, is illustrated in Figure 27-9. Note that special transformers are required.

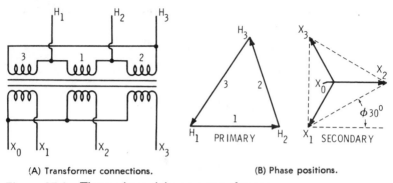

(A) Transformer connections.   (B) Phase positions.

Figure 27-8 Three-phase delta-wye transformer.

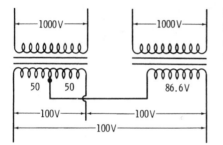

Figure 27-9 Voltage relationships in Scott-connected transformers for two phases to three phases.

The high-voltage windings usually have taps for adjusting the low-voltage output to the desired voltage if the incoming line voltage is either too high or too low. For instance, if it is desired to get 120/208 V output with a normal input of 480 V, this is one thing, but if the input is higher or lower than 480 volts, these taps will adjust for the desired voltage output.

For example, such transformers usually have taps in the vicinity of 2½% ranges, which may be internally changed, such as 456/467/480/504 volts. A diagram is shown on the nameplate of the transformer and the necessary changes are made to meet the voltage conditions.

Large transformers usually have an external dial for tap changing. Others may even have motor-operated tap changers, which change under load as the voltages change. These, of course, would be automatic.

Delta- and Wye-Connection Voltages and Currents

A comparison of winding and line voltages and current is shown in Figure 27-10 for delta connections and in Figure 27-11 for wye connections. Note that in Figure 27-10, for the delta connection, that the winding voltage and the line voltage are the same, while the line amperes are 1.73 times the winding current.

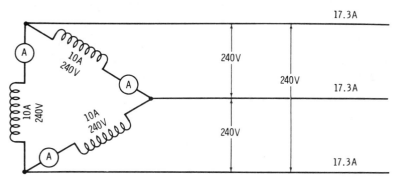

Figure 27-10 Voltage and current in a delta transformer.

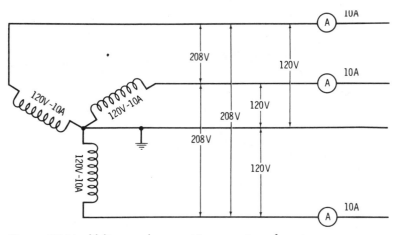

Figure 27-11 Voltage and current in a wye transformer.

In Figure 27-11, for the wye connection, the line voltage is 1.73 times the winding voltage, and the line current is the same as the winding current.

The above descriptions, of course, assume balanced loads and are representative only.

Questions

1. Sketch a schematic of three single-phase transformers connected in delta-delta.
2. Sketch a schematic of an open-delta connection.
3. Add a neutral connection to the drawing for question No. 2.
4. Sketch a schematic of a three-phase, delta-delta transformer.
5. Sketch a schematic of a three-phase wye-wye transformer.
6. What problems are encountered in wye-wye connections?
7. Sketch a schematic of a four-wire delta-wye.
8. Sketch a Scott connection. Tell what it is used for, and describe it.

Chapter 28

Autotransformers

Autotransformers differ from conventional transformers in that they have only one continuous winding. There are many *NEC* restrictions to their use, but they have a place in the electrical industry.

Autotransformers and Conventional Transformers

In the conventional transformer, the high-voltage winding and the low-voltage winding are electrically isolated. Thus, they are termed *isolation transformers*.

Figure 28-1 illustrates the type of transformer covered in Chapter 27 and gives voltages and current for this particular transformer. An autotransformer of the same capacity and voltage is illustrated in Figure 28-2. Particularly note that the two windings are as one and not isolated.

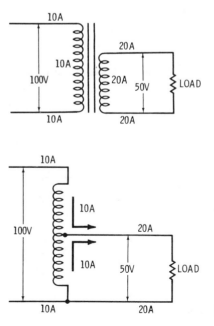

Figure 28-1 Ordinary transformer with voltages and current shown.

Figure 28-2 Autotransformer.

Figure 28-3 illustrates what happens to the output voltage if the connections are not made correctly. Figure 28-3A shows the right connections, and Figure 28-3B shows the connections to *G-E* of the winding reversed. As a result there is zero voltage at *C-D*.

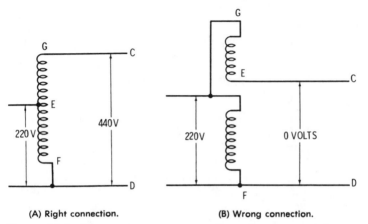

(A) Right connection.	(B) Wrong connection.

Figure 28-3   Right and wrong connections of an autotransformer.

Induction Voltage Regulator

The reversal of a part of the winding is shown to clarify the operation of an induction voltage regulator. These regulators are used as voltage boosters on long power lines and work automatically as the load increases or decreases to take care of voltage drops at the load caused by *IR* loss in the conductors. (See Figure 28-4.) They have built into them $RX_L X_C$ networks, which may be set to conform to

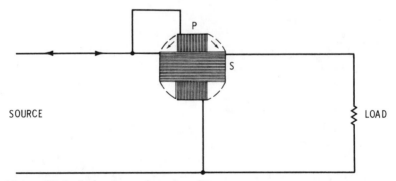

Figure 28-4   Induction voltage regulator.

the characteristics of the power line and thus automatically compensate for changes of load at the load end of the line and keep a stable voltage there.

The induction regulator consists of two windings. The secondary, *S*, is connected in series with the load. The primary, *P*, is connected in shunt across the line and is mounted so it will turn so that its inductive relationship with the secondary may be varied. The primary is the moving coil, because it is wound with finer, more flexible conductors. A regulator is an autotransformer with a variable ratio of transformation.

If the primary is rotated 90° in one direction, the line voltage will be boosted by the amount of emf induced into the secondary. If the primary is rotated 90° in the other direction, the voltage of the line will be lowered to the extent of the induced emf in the secondary. Thus, the amount of rotation and the direction of rotation, when controlled automatically, will compensate for the voltage drops in the load side of the line from the induction regulator. A phantom impedance circuit is built-in to simulate the line characteristics.

Transformer Booster

The conventional type of transformer is sometimes used on power lines to boost or buck line voltage, as an emergency measure. Great care should be taken in connecting it up, as will be explained.

Figure 28-5 shows such a transformer booster. The high-side (*A*) winding is for 1000 volts, and the low side (*B*) is for 100 volts. These are used as merely representative figures. The values used will depend on the line voltage and the amount of voltage boost required.

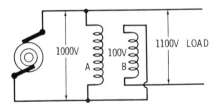

Figure 28-5 Transformer booster connection.

This connection will act as a current transformer, which will be explained in the next chapter. A word of caution, however, is necessary. Unless handled properly, the connecting of the booster may be extremely hazardous. A very high voltage may be produced in winding *A* if it is not connected to the line.

Thus, winding A *must* be connected to the line before winding B is connected and, also, winding A must be connected solid to the line and no fuses inserted in this connection. After winding A is connected, then winding B may be connected in series with the line, so as to add to the line voltage. See the material on checking polarity in Chapter 27. Winding B must be large enough to handle full line current without damage. Therefore, the full-load current of the low side of the transformer must be checked with the anticipated full-load line current, and the high-side voltage rating of this transformer must be that of the line being connected to.

Questions

1. Describe an autotransformer and compare it to a conventional transformer.

2. Sketch the connections for an autotransformer.

3. What is an induction regulator, and how does it work?

4. Sketch and describe fully a booster transformer and its connections.

Chapter 29

Instrument Transformers

In Chapter 14, voltmeters, ammeters, etc., were covered. It is not practical to use coil-wound voltmeters with resistance in series on high-voltage AC circuits, because of the excessive insulation that would be required and the large amount of resistance that it would take.

It is likewise not practical to use shunts and millivoltmeters for measuring currents on high-voltage AC circuits. In both instances, zero-temperature-coefficient resistors could be used, but they would have reactance changes with any change of frequency.

Use of Instrument Transformers

Instrument transformers are the practical answer for measuring current and voltages where high AC voltages and currents are encountered. There are two chief reasons for this:

1. Electricians and station operators are protected from contact with high voltage.

2. The instruments may be constructed with reasonable amounts of insulation and current-carrying capacity.

Instrument transformers are not only used for current and voltage measurements, but also for watt-hour meters, ground relays, overcurrent relays, synchroscopes, ground fault interrupters, and other uses that require their operation by high currents and/or high voltages.

Types of Instrument Transformers

Instrument transformers supply low currents and low voltages for the voltmeters, ammeters, etc. These transformers have very high accuracy of transformation and constants (K) are used as multipliers to convert the readings on the voltmeter or ammeter to the value of the actual voltage or current of the line or equipment being measured. Current transformers are usually designed for 5 amperes output maximum on the secondary, and potential transformers are usually designed for 120 volts output on the secondary.

In order to secure the greatest accuracy, both types of transformers must be made of the highest quality of core steel available, and precision of winding is also necessary.

The potential transformer is an isolation type of transformer, such as was just covered in Chapters 26, 27, and 28. That is, the primary winding is in shunt with the load, and the secondary is in shunt with the instrument(s) used in the measuring. See Figure 29-1. Potential transformers are often called PTs.

The current transformer is also an isolation-type transformer, but the primary is the line conductor or bus; or it may be more than one conductor, that is, two or more turns of conductors capable of handling the total line current. The primary winding is in series with the load, instead of in shunt as with the potential transformers. The primary current is thus the load current, and the primary emf is the drop of potential across the transformer due to its primary impedance. These transformers are commonly called CTs. Figure 29-1 illustrates a CT as indicated by C.

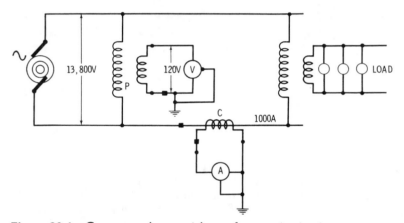

Figure 29-1 Current and potential transformers in circuit.

Potential Transformers

The primary emf of a potential transformer is the line voltage of the circuit being measured. In Figure 29-1, the alternator supplies 13,800 volts to a line. Potential transformer P steps the voltage down to 120 volts for measurement. This is a ratio of 115 to one, so K for the potential transformer is $K = 115$. This indicates that the reading on a 120-volt meter has to be multiplied by 115 to obtain the value of the primary voltage.

This sketch shows 1000 A in the line. Assume this to be the maximum value of the CT. Then $1000/5 = 200$, or $K = 200$ for the current transformer. This indicates that the ammeter reading must be multiplied by 200 to get the line current.

The scales on the meter may be marked to take the K into account in the direct reading on the scale, or the reading may have to be multiplied by K.

Current Transformers

Figure 29-2 shows the arrangement of a current transformer. This is the doughnut type, which is used extensively at service entrances for measuring high current capacities.

In Chapter 14 a picture of an Amprobe clamp-on ammeter was shown (Figure 14-9). The difference between the CT in Figure 29-2 and the clamp-on ammeter is that the laminated iron core, C, is hinged so that it may be opened for clamping around a conductor to avoid having to open up a circuit to use the ammeter.

In Figure 29-2, C is a laminated iron core and B is the line conductor or bus and also the primary winding of the CT. S is the secondary winding, made up of a number of turns of small-size conductors, and A is an ammeter.

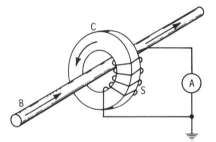

Figure 29-2 Theoretical arrangement of a current transformer.

Ratings

CTs are usually rated with the voltage and the maximum amperes anticipated on the conductors with which they are to be used. There is also a rating, such as 100 to 5, 200 to 5, 1000 to 5, etc. This means they are rated 100 amperes in the line conductor to 5 amperes on the meter, etc.

Opening the Secondary Circuit

The secondary circuit of a CT *must never be opened* when under load. Current transformers are often designed so that the cover over the secondary terminals shorts the secondary out if the cover is removed.

There are two reasons for never opening the secondary under load:

1. A dangerously high voltage may be developed that may cause injury, even if not fatal, should anyone come in contact with the secondary circuit.

2. The high flux resulting from the open circuiting of the secondary may saturate the iron to such a degree as to seriously impair the ratio of the transformer, making it unfit for further use where accurate power measurements are required.

Consider the opening of the secondary of a CT under load conditions by comparing it with a conventional potential transformer. If the secondary load of a conventional potential transformer is removed, the current decreases in the primary until only the exciting current remains. In a current transformer the primary current is always the current in the line conductor, so when the secondary of a CT is opened, all of the primary current becomes exciting current. The flux and the induced voltage in the primary may rise to a high value. This abnormal flux will raise the secondary voltage to a very high value, depending upon the amount of current in the line conductor. It may puncture the secondary insulation or produce an arc at the point where the secondary circuit has been opened.

Symbols

The usual symbols for potential transformers are shown in Figure 29-3. The symbols for current transformers are shown in Figure 29-4.

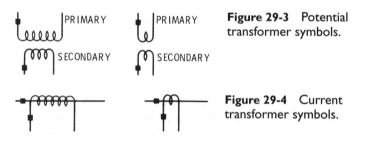

PRIMARY PRIMARY **Figure 29-3** Potential transformer symbols.

SECONDARY SECONDARY

Figure 29-4 Current transformer symbols.

Ground Fault Protection

Current transformers are used for ground fault interrupters (GFIs) and ground fault circuit interrupters (GFCIs). These two are frequently referred to as zero-sequence transformers.

In the chapter on electromagnetism, it was shown that if all conductors of a circuit were kept together, the magnetic flux encircling the conductors cancelled out. This is the principle of the operation of ground fault circuit interrupters.

Figure 29-5 illustrates the principle involved. The current transformer, CT, encircles all of the conductors, A, B, C, and N. When there is no ground fault on the load side of the transformer, the flux cancels out so no current flows from the secondary through the relay coil.

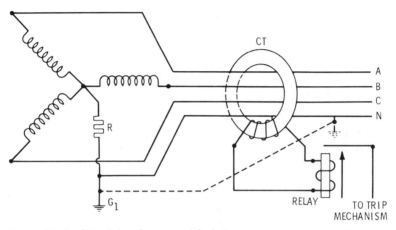

Figure 29-5 Principle of a ground fault interrupter.

Now a ground (G_2) occurs on phase C, and a part of the current in C returns through the equipment-grounding conductor or the earth, by way of the dashed line. Thus, there occurs an unbalance in the circuits through the CT coil, causing a flux that induces a voltage into the secondary, S. Current flows through the relay, energizing it and closing the contacts, which causes the circuit to be tripped off.

The *NEC* requires GFIs on high-capacity services to prevent burn-downs of equipment due to faults.

Figure 29-5 is the simplest form. There are many variations and uses. One is to sound an alarm to indicate that one phase is grounded. Note resistor R in the neutral ground. This resistor type of neutral grounding prevents shutdowns when a ground occurs. If two phases ground, then the overcurrent devices will trip or blow.

Another such device is the GFCI, which may be called the people protector. It is called for on receptacle circuits and other circuits in many places in the *NEC*. Figure 29-6 shows a GFCI.

The GFCI is set to trip at 0.005 ampere. If the hot conductor goes to the case of a saw or other equipment and the grounding conductor is intact and properly connected, the fuse should blow or the breaker trip. If one were holding the saw, for example, and

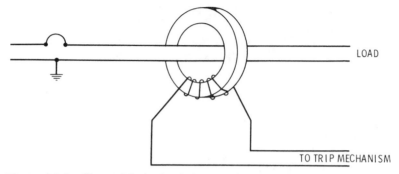

Figure 29-6   Ground fault circuit interrupter.

standing on the ground, in all probability one wouldn't get hurt. If the ground were of high resistance, however, the breaker might not trip and the person might be killed.

With the GFCI in the circuit, as soon as 5 milliamperes passed through the person to ground, the secondary would open instantly. The person would receive a shock but would live to tell about it.

GFCIs are available in separate portable equipment, receptacles, both plug-in and permanent types, and also in circuit breakers for panel installation.

Questions

1. Why are instrument transformers used? Explain fully.

2. What is the most common maximum secondary current of a current transformer?

3. What is the most common secondary voltage of a potential transformer?

4. Explain how the constant is figured for a current transformer.

5. Explain how the constant is figured for a potential transformer.

6. Give the symbols for instrument transformers and label them.

7. Should the secondary of a current transformer ever be opened under load? Explain.

8. What does GFCI mean?

9. Explain the operation of a GFCI.

10. At what current value are people-protector GFCIs set for?

Chapter 30

Polyphase Induction Motors

Basically, there are three general types of AC motors, namely

1. Synchronous motors
2. Polyphase induction motors
3. Single-phase motors

This chapter will deal with polyphase induction motors, as this type of motor is the most used of the three types and the most trouble-free. Also, the theory that will be learned here will fit in very well with that of the other two types of motors.

The polyphase induction type of motor depends upon the principle of a rotary magnetic field. Polyphase motors using a rotating magnetic field were invented by Nikola Tesla in 1898.

Rotating Magnetic Field

To create a rotary magnetic field of force, it is necessary to have two or more magnetomotive forces acting to produce a flux in the same motor field area, but at a phase angle displacement of both time and space with respect to each other. The two or more magnetic fields combine to create a resultant magnetic field that may be rotated in both directions, and the speed of rotation depends upon the frequency of the AC supply.

Figure 30-1 illustrates the basic principle of a two-phase rotating field, with a phase relationship of 90°. The two-phase field is used for simplification. Three-phase rotating fields produce similar results, except there are three fields of force 120° apart, instead of 90° apart. (In Chapter 23, covering polyphase systems, it was shown that with two-phase systems, the displacement between phases is 90°.)

In Figure 30-1, phase 1 is at a maximum position, 1′; thus, pole A is north and pole $A′$ is south, and phase 2 is zero. Thus, the compass needle points north (N). The poles covered by phase 1 are A and $A′$, and the poles covered by phase 2 are B and $B′$. For position 2′, phase 1 and phase 2 are both in a partial positive direction; thus, poles A and $B′$ are both north and poles B and $A′$ are both south, and the compass needle has moved 45° clockwise. In position 3′, phase 1 is zero and phase 2 is maximum positive; thus, we

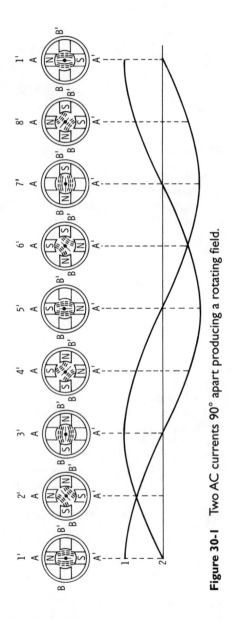

Figure 30-1 Two AC currents 90° apart producing a rotating field.

have pole *B'* north, and pole *B* is south, and the compass needle has moved another 90° clockwise.

By following the two sine waves, magnetic poles, and positioning of the compass needle, it will be seen that as phases 1 and 2 move 360°, the compass has made one 360° movement from 1' back to 1'. Thus, the magnetic field has rotated 360° clockwise in one cycle.

Since most motors are three-phase, they will have three separate windings per pole on the stator (outside stationary portion of the motor). These three pole windings and their currents are 120° apart, and as they rotate around the stator winding, a rotating field is created. The direction of rotation depends upon the phase connections to the stator winding. If a three-phase motor runs the opposite direction to what is wanted, you merely have to reverse the connections of two phase conductors of any of the three conductors. This will reverse the direction of field rotation and the direction of the rotor rotation.

The speed of the rotating magnetic field depends upon the number of poles for which the motor is wound and the frequency and number of phases of the applied AC current. The speed referred to here is the synchronous speed, and the following formula is used to determine the synchronous speed:

$$\text{Synchronous Speed in rpm} = \frac{120 \times \text{Frequency in Hertz}}{\text{Number of Poles}}$$

$$S = \frac{120f}{P}$$

where

S = synchronous speed in revolutions per minute
f = frequency in hertz
P = number of poles

As an example, a three-phase four-pole motor at 60 Hz has a synchronous speed of 1800 rpm, which is calculated as follows:

$$S = \frac{120 \times 60}{4} = 1800$$

A three-phase six-pole motor at 60 Hz has a synchronous speed of 1200 rpm:

$$S = \frac{120 \times 60}{6} = 1200$$

The stator or stationary part of the motor has the winding to which the three-phase voltage is applied. The rotor or moving portion of the motor has a much different type of winding from the stator, and the rotor winding is in no manner connected to the applied voltage. It receives its voltage by induction, as does a transformer.

Squirrel Cage Rotor Winding

Figure 30-2 illustrates the squirrel cage winding on a rotor. It is composed of metal bars or rods through holes in the rotor laminations, and these rods or bars are welded, brazed, die-cast, etc., to metal rings around the outside of the motor rotor. This in reality constitutes metal bars through the laminations, and these bars are short-circuited by the metal and rings.

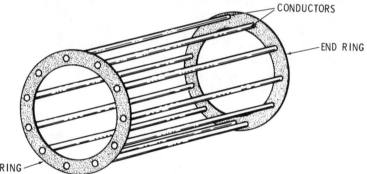

CONDUCTORS

END RING

END RING

Figure 30-2 Squirrel cage winding.

In the comparison just made to transformers, the stator winding is the primary winding, and the rotor squirrel cage is the secondary winding, but it is a short-circuited winding.

The theory of the manner in which current is induced in the rotor may be illustrated by Figure 30-3. Suppose that a rotating field of force, due to currents in the stator windings, rotates from polar projections A to B to C to D, in the direction of the arrow, K. The lines of force due to this field are in direction D-B. As the field rotates, it cuts squarely across conductors E-F and G-H of the rotor, which for the moment are stationary. This direction of cutting will induce currents that will flow toward the observer in the conductors E-F and away from the observer in conductors G-H. The direction of the resultant flux of these induced currents will evidently be A-C. This produces a north pole on the rotor at N, which would be attracted to the south pole of the stator, S', while the north pole of the rotor, N, would be repelled by the north pole of the stator, N', causing the rotor to move in the direction K of the rotating field. If the rotor traveled as fast as the rotating field, there would be no induced voltage or current in the rotor winding and therefore no torque, because in order to induce a voltage in the rotor winding, there must be relative motion between the rotor and rotating field. Thus, it is essential that the rotor run at a speed below synchronous speed. The difference between synchronous speed and the speed of the rotor is called *slip*. This slip usually is in the range of 2 percent to 5 percent.

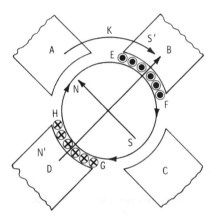

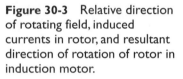

Figure 30-3 Relative direction of rotating field, induced currents in rotor, and resultant direction of rotation of rotor in induction motor.

Slip is responsible for the regulation of a polyphase induction motor. At no load the rotor runs at near synchronous speed. As the load is applied to the motor, the slip increases, causing more induced voltage, which results in more current in the rotor winding to compensate for the additional load; and, as with a transformer, more current is drawn by the stator (primary).

The torque increases with the slip up to a point. If the rotor stands still (locked rotor) while the rotating field travels 1800 rpm due to a frequency of 60 Hz on a three-phase four-pole motor, the frequency induced into each conductor of the rotor would also be 60 Hz. With no load, the motor will run at near synchronous speed. There would be very little slip and the frequency of the current induced in the rotor would be near zero. If load were added so as to cause the rotor to slip 5 percent below synchronous speed, the frequency of the currents induced in the rotor would be 5 percent of 60 Hz, or 3 Hz.

If the rotor possessed no inductance but only resistance, the concept of magnetic polarity of the field and the current in the rotor may be gained from Figure 30-4. The induced voltage and resultant current in the rotor would be maximum directly under the center of the stator poles. These poles would react with a maximum effect upon the poles of the rotating field to produce torque.

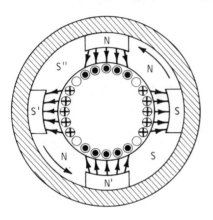

Figure 30-4 Magnitude and position of currents of induction motor if rotor possesses resistance only.

A resistive rotor without inductance is hypothetical. If the winding of a rotor possesses great inductance and comparatively little resistance—so that the currents wouldn't reach their maximum until the rotor conductors reached a point such as illustrated in Figure 30-5, where the resultant polarity of the rotor poles would be in line with the poles of the stator—the motor torque would be zero.

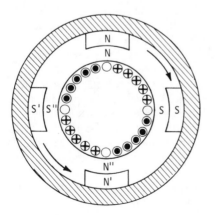

Figure 30-5 Magnitude and position of currents of induction motor if rotor possesses inductance only.

From this, it may be said that at start,

1. The frequency of the current in the rotor is maximum.

2. The reactance of the rotor winding is great.

3. The torque is small.

As the rotor comes up to speed, the frequency of the current in the rotor becomes less; thus, the reactance of the rotor becomes less; and the angular displacement of the poles of the rotor and poles of the stator is less; and the condition in Figure 30-4 would be approached and the torque increase.

To recap, at synchronous speed there would be no slip of the rotor, hence no current and no torque. Therefore, there must be some slip. As slip increases, so does the rotor current, and the result is increased torque. The slip and torque don't increase proportionally because of the demagnetizing effect of the rotor currents.

The starting torque of an induction motor depends largely upon the rotor resistance. A low-resistance rotor will have poor starting torque and high starting current. A high-resistance rotor will have high starting torque and low starting current. A high-resistance rotor will have poorer speed regulation than a low-resistance rotor.

Figure 30-6 illustrates a representative torque characteristic curve for an induction motor. The rotor bars and slots should never

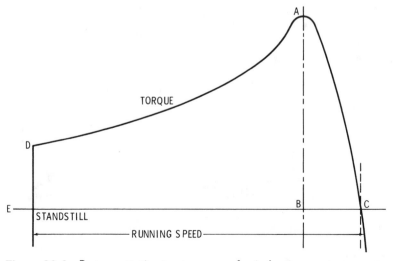
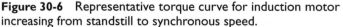

Figure 30-6 Representative torque curve for induction motor increasing from standstill to synchronous speed.

be equal to or a multiple of the number of slots in the stator (Figure 30-7).

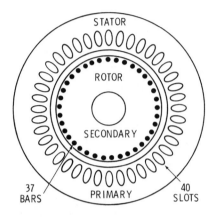

Figure 30-7 Stator slots and rotor bars should not be equal to or multiples of each other.

Wound Rotors

A polyphase induction motor need not have a squirrel cage rotor winding, but may have a wound rotor with slip rings and an external adjustable rheostat to be used in starting. At start, the resistance is cut into circuit, making a high-resistance rotor. As the speed picks up, the resistance is gradually taken out of circuit, until the rotor winding is shorted out across the slip ring (Figure 30-8).

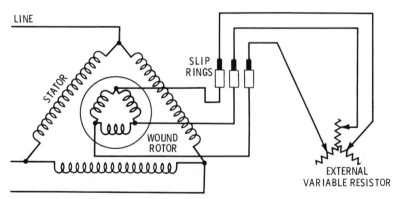

Figure 30-8 Induction motor with wound rotor.

Dual Voltages

Three-phase induction motors are usually wound for dual voltages, such as 240 volts and 480 volts, etc. The leads out of the motor are numbered and connected for the voltage desired. This is accomplished by connecting internal windings in series or in parallel. The internal connections of the motor may be connected wye or delta, but wye is the most used.

A schematic of a three-phase squirrel cage induction motor is shown in Figure 30-9. This is delta-connected. The internal and external connections for a dual-voltage, delta-wound motor are illustrated in Figure 30-10.

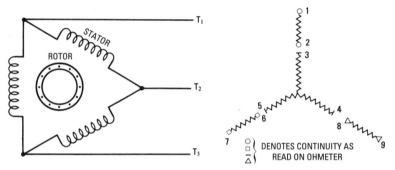

Figure 30-9 Delta-connected induction motor.

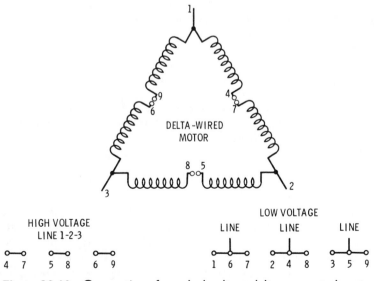

Figure 30-10 Connections for a dual-voltage delta-connected motor.

Figure 30-11 illustrates the internal windings and external connections for a dual-voltage wye-wound motor.

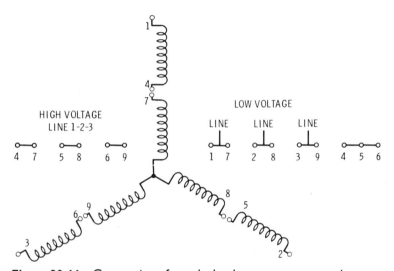

Figure 30-11 Connections for a dual-voltage wye-connected motor.

An easy way to remember how to connect the leads of a dual-voltage motor, connected for wye, is shown in Figure 30-12. First draw the arrangements of wye windings. Then, starting at one point, draw a spiral so that it connects in order with all six coils. Begin at the starting point and number from 1 through 9 as you progress around the spiral. From this it is easy to see how to series or parallel coils as needed.

The following description is a method of determining the correct wiring sequence of a three-phase motor that for some reason doesn't have the leads from the motor numbered. You don't need to disassemble to find the proper numbering of the motor leads.

Step 1: Number the nine leads arbitrarily so that you have a starting point.

Step 2: With an ohmmeter or similar device, find the various coil sets and make a record of these coil sets, using the numbers that you originally assigned to the leads; assuming the motor is wye-connected, only three leads will show continuity to all of the thru leads. These three leads are the internally

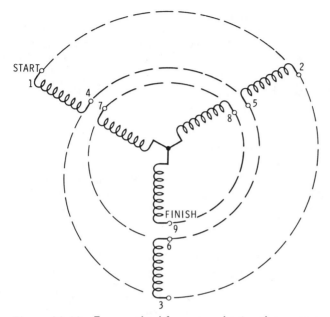

Figure 30-12 Easy method for remembering the proper connections for a dual-voltage wye-connected motor.

connected wye. The remaining six leads belong to the three isolated sets of coils. See Figure 30-9.

Step 3: If the motor is Y connected, follow this procedure:

(*a*) Apply reduced voltage (40–100 volts) across one of the sets of coils that is in the internal, not in the wye. Use Figure 30-9A as an example, and put voltage across 6 and 7. Read the voltages across the internal wye that you have located (from 3 to 4, 4 to 5, and 3 to 5). Pick the lowest of these voltages. (Let us assume that this is between 3 and 4.)

(*b*) The terminal of the wye that the voltmeter is not attached to when the lowest voltage is read is the terminal that should be connected to the energized coil. In the example, since the voltmeter is connected across 3 and 4 when the lowest voltage is read, terminal 5 should be connected to coil 6 or 7.

(*c*) It now remains to find out whether 6 or 7 should be connected to 5 for correct operation of the motor. First connect 5 to 6 and apply reduced voltage to 3 and 7. Read voltage across 6 and 7. Now connect 7 to 5 and apply voltage to 3 and 6. Read the voltage across 6 and 7. The connection that resulted in the highest voltage across 6 and 7 is the correct connection.

Step 4: Repeat the above for the other coil sets. Remove the numbers you originally put on the leads and renumber them properly as shown in Figure 30-11.

Induction Motors on Start

Induction motors on start draw far in excess of full-load current. This current is termed *locked-rotor current,* and the motor nameplate has a code letter to indicate the locked-rotor current. The handiest method of interpreting the code letter is to use Table 430.7(b) of the *NEC*. This table is reproduced here as Table 30-1.

Table 30-1   [Table 430.7(b) of the NEC] Locked-Rotor Indicating Code Letters*

Code Letter	Kilovolt-Amperes per Horsepower with Locked Rotor
A	0–3.14
B	3.15–3.54
C	3.55–3.99
D	4.0–4.49
E	4.5–4.99
F	5.0–5.59
G	5.6–6.29
H	6.3–7.09
J	7.1–7.99
K	8.0–8.99
L	9.0–9.99
M	10.0–11.19
N	11.2–12.49
P	12.5–13.99
R	14.0–15.99
S	16.0–17.99
T	18.0–19.99
U	20.0–22.39
V	22.4–and up

* *This table is an adopted standard of the National Electrical Manufacturers' Association*

For a motor with code letter G, the range is 5.6 to 6.29 kVA. Thus for a 20-hp motor, the actual kilovolt-ampere range would be 20×5.6 to $20 \times 6.29 = 112$ kVA to 125.8 kVA for locked rotor, which would also be the current drawn at standstill as the motor was first connected across the line for starting.

Assume this 20-hp motor to be three-phase, 230 volts, 54 amperes full-load current. Then the locked-rotor current can be calculated thus:

$$\text{Lower Limit} = \frac{112,000}{1.732 \times 230} = 281 \text{ amperes}$$

$$\text{Upper Limit} = \frac{125,800}{1.732 \times 230} = 316 \text{ amperes}$$

where $1.732 = \sqrt{3}$ is a constant applying to three-phase motors. The locked-rotor current at start would be instantaneous; with the first motion, it would begin to reduce in value.

Many three-phase induction motors are started by connecting them across the line. Some utility companies set upper limits as to the sizes that may be so started. If a motor is not permitted to start across the line, there are several methods of reducing starting current:

1. Start with the winding connected wye, which in essence series the windings. Then, after the motor gets partway up to speed, the starter automatically cuts the winding over to a delta-run connection. These starters usually use current transformers and current relays for the changeover.

2. By using starting compensators, which use autotransformers to reduce the voltage to the motor at start and increase available current to the motor, while reducing the line current drawn. Figure 30-13 is a schematic for a starting compensator.

The starting contacts and the contacts for the autotransformer close at start, and the motor receives reduced voltage and increased current. After it begins to gain sufficient speed, the starting contacts and the autotransformer contacts open and the running contacts close, which connects the motor across the line and cuts the autotransformer out of circuit.

This may be accomplished by means of hand control or may be done automatically. The transformer usually has several taps on it, so the starting current may be set at the value desired.

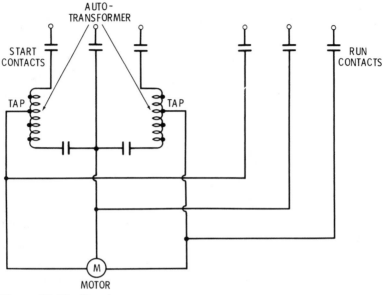

Figure 30-13 Starting compensator.

It is not the intent of this book to cover all types of motor controls, as this is a subject unto itself, but a standard diagram (Figure 30-14) and standard schematic diagrams (Figures 30.15 and 30-16) of across-the-line magnetic starters are shown.

Altitude Correction

Mention must be made of the necessity of making altitude corrections on motors. Motors operating from sea level to 3300 ft need not be derated, but above this altitude the output of the motor must be derated 0.3 percent for each 330 ft over 3300 ft, or 1 percent for each 1000 ft above 3300 ft. This is necessitated by the decrease in air density with altitude and the resultant decrease in the cooling effect of the air.

Power Factor

Induction motors are basically an inductive reactance load, and as a result will cause a lagging power factor. At no load, this lagging power factor is the greatest, and at full load the lagging power factor decreases. This is brought to your attention because the induction motor is one of the greatest contributors to a lagging power factor.

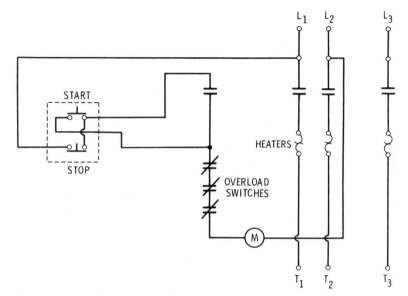

Figure 30-14   Diagram of a three-phase magnetic starter.

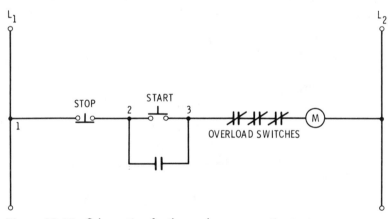

Figure 30-15   Schematic of a three-phase magnetic starter.

It would be amiss to close this chapter without stating that if a squirrel cage induction motor is driven by a prime mover at above synchronous speed, it will become an alternator if connected to another source of AC current, and will deliver power, but only power at 100 percent power factor. It is also good to recognize that

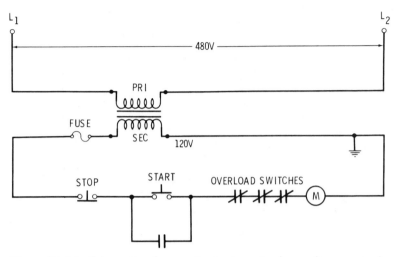

Figure 30-16   Schematic of magnetic starter using low-voltage controls.

rotating equipment will add to available fault currents for a very short time, and this must be considered in figuring available fault currents. The usual value to add to available fault currents is five times the full-load current.

Questions

1. Name three general types of AC motors.
2. Illustrate and explain what a rotating field is.
3. Give the formula for synchronous speed.
4. Describe a squirrel cage rotor.
5. What is slip? Explain fully.
6. Describe how a squirrel cage motor gets its regulation.
7. Explain how torque and rotor resistance affect motor starting.
8. Sketch and explain a wound-rotor motor.
9. Explain locked-rotor current.
10. Name and explain three methods of connecting a squirrel cage motor for starting.
11. Explain altitude correction for motors.

Chapter 31

Power-Factor Correction

The bad effects of poor power factor have been covered. It seems appropriate that the method of correcting power factor be given: that is, how to calculate capacitive reactance to correct lagging power factor caused by inductive reactance.

Reactive Kilovolt-Amperes

Kvars are used in expressing the values of capacitive reactance involved in power factor correction. These are called *kilovars* and are reactive kilovolt-amperes.

Most of the reactive loads that cause power factor problems consist of inductive reactance, causing lagging power factor. Thus, when kvars are mentioned, it may be assumed that reference is being made to capacitive reactance.

The kilowatt value of the load involved in the circuit must be determined by calculation or by the use of a wattmeter.

Use of Correction Table

Table 31-1 is a standard table used in calculating kilovars for power factor correction. This table gives the original power factor in percent in the left-hand column and the desired power factor at the top of the table (this is the power factor to which we want to bring the circuit).

In explanation of how to use this table, the following example is given: A 50-hp, three-phase motor draws 65 amperes at 460 volts and the power factor is found to be 75 percent. It is desired to bring the power factor of this motor circuit up to 95 percent:

$$460 \text{ V} \times 65 \text{ A} \times 1.73 = 51,727 \text{ VA}$$

$$51,727 \text{ VA @ } 57\% \text{PF} = 51,727 \times 0.75 = 38,795.25 \text{ watts}$$

In the left-hand column, find 75, which is the original power factor. At the top, find 95, which is the PF the circuit is to be corrected to. Using a straightedge, lay it horizontally under the 75 and follow over to the column under 95. The constant at this point is found to be 0.553. Change 38,795.25 watts to kW: 38,795.25W/1000 = 38.795 kW. Multiply the kilowatts by the constant: 38.795 × 0.553 = 21.45 kvars.

Table 31-1 Capacitor Kvar Table for Improving Power Factor—Desired Power Factor in Percent

%	Original Power Factor in Percent															
	80	**81**	**82**	**83**	**84**	**85**	**86**	**87**	**88**	**89**	**90**	**91**	**92**	**93**	**94**	**95**
50	.982	1.008	1.034	1.060	1.086	1.112	1.139	1.165	1.192	1.220	1.248	1.276	1.303	1.337	1.369	1.402
51	.936	.962	.988	1.014	1.040	1.066	1.093	1.119	1.146	1.174	1.202	1.230	1.257	1.291	1.320	1.357
52	.894	.920	.946	.972	.998	1.024	1.051	1.077	1.104	1.132	1.160	1.188	1.215	1.249	1.281	1.315
53	.850	.876	.902	.928	.954	.980	1.007	1.033	1.060	1.088	1.116	1.144	1.171	1.205	1.237	1.271
54	.809	.835	.861	.887	.913	.939	.966	.992	1.019	1.047	1.075	1.103	1.130	1.164	1.196	1.230
55	.769	.795	.821	.847	.873	.899	.926	.952	.979	1.007	1.035	1.063	1.090	1.124	1.156	1.190
56	.730	.756	.782	.808	.834	.860	.887	.913	.940	.968	.996	1.024	1.051	1.085	1.117	1.151
57	.692	.718	.744	.770	.796	.822	.849	.875	.902	.930	.958	.986	1.013	1.047	1.079	1.113
58	.655	.681	.707	.733	.759	.785	.812	.838	.865	.893	.921	.949	.976	1.010	1.042	1.076
59	.618	.644	.670	.696	.722	.748	.775	.801	.828	.856	.884	.912	.939	.973	1.005	1.039
60	.584	.610	.636	.662	.688	.714	.741	.767	.794	.822	.849	.878	.905	.939	.971	1.005
61	.549	.575	.601	.627	.653	.679	.706	.732	.759	.787	.815	.843	.870	.904	.936	.970
62	.515	.541	.567	.593	.619	.645	.672	.698	.725	.753	.781	.809	.836	.870	.902	.936
63	.483	.509	.535	.561	.587	.613	.640	.666	.693	.721	.749	.777	.804	.838	.870	.904
64	.450	.476	.502	.528	.554	.580	.607	.633	.660	.688	.716	.744	.771	.805	.837	.871
65	.419	.445	.471	.497	.523	.549	.576	.602	.629	.657	.685	.713	.740	.774	.806	.840
66	.388	.414	.440	.466	.492	.518	.545	.571	.598	.626	.654	.682	.709	.743	.775	.809
67	.358	.384	.410	.436	.462	.488	.515	.541	.568	.596	.624	.652	.679	.713	.745	.779
68	.329	.355	.381	.407	.433	.459	.486	.512	.539	.567	.595	.623	.650	.684	.716	.750
69	.299	.325	.351	.377	.403	.429	.456	.482	.509	.537	.565	.593	.620	.654	.686	.720
70	.270	.296	.322	.348	.374	.400	.427	.453	.480	.508	.536	.564	.591	.625	.657	.691
71	.242	.268	.294	.320	.346	.372	.399	.425	.452	.480	.508	.536	.563	.597	.629	.663
72	.213	.239	.265	.291	.317	.343	.370	.396	.423	.451	.479	.507	.534	.568	.600	.634
73	.186	.212	.238	.264	.290	.316	.343	.369	.396	.424	.452	.480	.507	.541	.573	.607
74	.159	.185	.211	.237	.263	.289	.316	.342	.369	.397	.425	.453	.480	.514	.546	.580
75	.132	.158	.184	.210	.236	.262	.289	.315	.342	.370	.398	.426	.453	.487	.519	.553

Correction with Capacitors

Capacitors for power-factor correction may be connected to the circuit in several different ways. Figure 31-1 illustrates capacitors connected across the motor load; Figure 31-2 illustrates capacitors connected ahead of the motor overload protection; and Figure 31-3 illustrates capacitors connected ahead of all motor loads. This could be at the service entrance to the building.

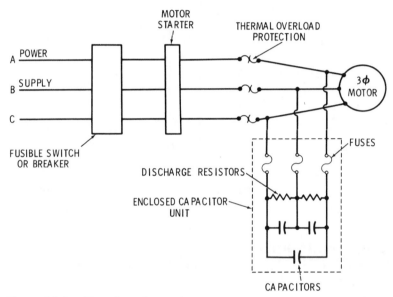

Figure 31-1 Capacitors located at motor.

Capacitors are required to be fused and have resistors across their terminals to bleed the charge of the capacitor away to keep from giving shocks. When the capacitor is connected at the motor, the motor will also assist in bleeding off the charge.

Use of Chart

Another method of calculating kvars is shown in Figure 31-4. Using the same problem as before, lay a straightedge from 75 on the left column to 95 in the right column. The straightedge crosses the middle column (percent reactive kVA) at 0.55 + or 0.553. This is the same as the constant found by using Table 31-1. Thus, $38.795 \times 0.553 = 21.45$ kvars., and the chart has been found to check out with the table.

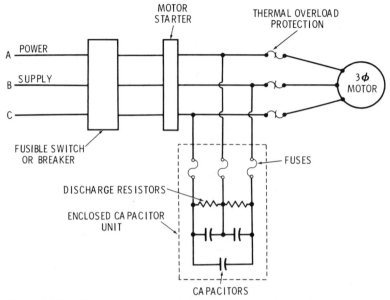

Figure 31-2 Capacitors located ahead of motor thermal overload protection.

Use of Mathematics

For those who enjoy mathematics, the kvars may be calculated another way.

Let

ϕ_a = original phase angle

ϕ_b = improved phase angle

PF_a = power factor before correction

PF_b = power factor after correction

$kvar_c$ = the rating of capacitor required to correct to PF_b

kW = kilowatt value of the load (true power)

Again, the same motor we discussed before will be used for the calculations. Angle ϕ_a may be arrived at from the power factor of 75 percent or cos ϕ_a = 0.75, and from the trigonometric tables, the

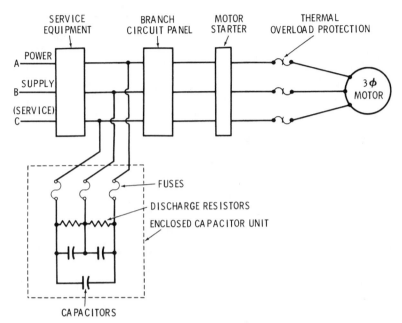

Figure 31-3   Capacitor located at service equipment to cover entire motor load.

angle is found to be 41°. Then the tangent of 41° will be found to be 0.869.

Again, ϕ_b may be arrived at in the same manner. Thus 95 percent = $\cos \phi_b$ = 0.95, and the angle is found to be 18°. Thus $\tan \phi_b$ is found to be 0.325.

$$\mathrm{kW} = \frac{460 \text{ V} \times 65 \text{ A} \times 1.73 \times 0.75}{1000} = 38.80 \text{ kVA}$$

$$\tan\phi_a = 0.869$$

$$\tan\phi_b = 0.325$$

$$\mathrm{kvars} = \mathrm{kW}(\tan\phi_a - \tan\phi_b) = 38.81\ (0.869 - 0.325)$$

$$= 21.10 \text{ kvars}$$

A motor with corrected power factor draws fewer amperes than the same motor without power factor correction.

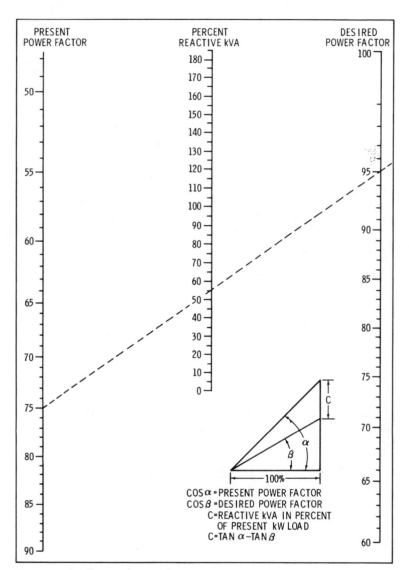

Figure 31-4   Power-factor chart.

Overloads

It was mentioned that the three-phase, 460-V, 50-hp motor was rated at 65 amperes at full load. The usual overload protection in the motor starter is 115 percent of full-load current. This value will vary with types of motors and their usage; see the *NEC*.

When corrected to 95 percent PF, the full-load current would be

$$460 \text{ V} \times I \times 1.73 \times 0.95 = 38.80 \text{ kVA}$$

or

$$756.01I = 38,800 \text{ VA}$$

Thus, $I = 38,800/756.01 = 51.3$ amperes. Theoretically, the overloads could be set at 51.3×1.15 or 59 amperes. If the capacitor is placed at the motor, the current drawn will be reduced and it would be possible to reduce the size of the overload protection. Good practice dictates that overload protection and conductor ampacity remain at the values they would be for uncorrected power factor.

Heating of the motor is based on full-load amperes. Therefore, if the overload protection remains at the value without power factor correction, should the capacitor be disconnected or the fuses to the capacitor blow, the motor wouldn't be bothered by nuisance trip-outs, and is still properly protected.

Reasons for Corrections

Very often, power factor correction is at the service entrance, as shown in Figure 31-3. The reasons for power factor correction are

1. Reduction in current drawn
2. Reduced voltage drop
3. Less slip in the motor
4. Less I^2R loss
5. Decreased transformer loss
6. Realization of rated kilowatt output capacity of transformers and generation equipment
7. Allows additional loads to be served without increased capacity
8. Lower power bills, where power factor penalties are imposed

Questions

1. Explain the meaning of kvars.
2. Is the average electrical load capacitive or inductive?
3. How may you ascertain the kilowatt value of a load?
4. A motor load draws 100 amperes at 460 volts and is found to have a power factor of 80 percent. It is desired to bring the power factor up to 95 percent. How many kvars of capacitive reactance will be required?
5. An inductive load causes a power factor.
6. A capacitive load causes a power factor.
7. Sketch three methods of connecting a capacitor into a motor circuit.

Chapter 32

Synchronous Motors

A synchronous motor is practically a duplicate of an alternator. (The alternator was covered in Chapter 17.) If an alternator is used as a synchronous motor, there must be an external driving force to bring it up to speed before it is connected to the electrical source of power. This is not especially practical, so other means of starting will be described later in this chapter.

Characteristics

Synchronous motors have a DC rotating field and a stationary AC winding or stator. The DC is supplied to the rotor by means of slip rings supplied from a DC source, such as an exciter on the motor shaft or from some other DC source.

Synchronous motors must run pole for pole with the alternator supplying the AC. This in most cases will be 60 Hz. Notice that we stated pole for pole and not the same rpm. The alternator may be an eight-pole machine, and the motor may be a four-pole machine. Thus, the motor would run twice as fast in rpm as the alternator is driven. Any variation in the speed on a pole-to-pole basis will therefore throw the two out of step and the motor would shut down.

Operation

In Figure 32-1, an AC current is delivered to the stator of a synchronous motor so that at a given instant the polarity produced by the current would be as shown, N and S. Consider a pivoted permanent magnet N'-S' between poles N and S. The polarity established in Figure 32-1 would cause repulsion of the permanent magnet N'-S', but would simply exert a direct repulsion without causing rotation.

If the magnet were turned to N"-S", there would be no improvement because with 60 Hz the stator poles change polarity every 1/120 of a second; thus, there would still be no motion. If the motor were caused to turn by an external source, and if, when it was nearly up to speed (synchronous), voltage was applied to the stator, the rotor would rotate and the external driving force could be removed.

The rotor turns pole for pole with the alternator, but there is a variable angle of lag between the motor rotor and the alternator

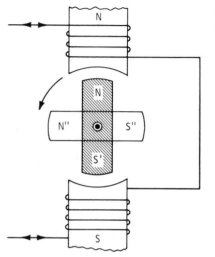

Figure 32-1 Principle of the synchronous motor.

rotor. This is illustrated in Figure 32-2, and, as may be seen, the attraction and repulsion of the rotor and stator poles will have a rotating effect. This angle of lag is what gives the synchronous motor its regulation.

The other means of starting a synchronous motor is to put a squirrel cage winding on the rotor in addition to the DC winding. Thus, it may be started as a squirrel cage induction motor. The motor DC field is not to be energized at start. After the squirrel

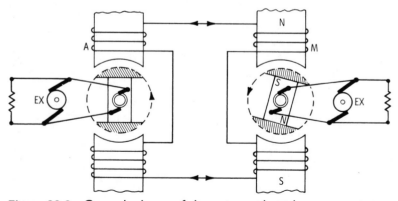

Figure 32-2 General scheme of alternator and synchronous motor combination with separate exciters.

cage brings the rotor up to a speed slightly below synchronous speed, the DC field circuit may be energized, the rotor pulls into step with the alternator, and the motor runs at synchronous speed.

Power Factor

The power factor of a synchronous motor may be varied by the excitation of the DC field. Figure 32-3 illustrates the operating curves of such a motor. As may be observed from these curves, an underexcited field causes a lagging power factor and also increases the AC stator current, as with all poor power factors. At normal excitation the power factor is 100% and the AC stator current is a minimum. Then, by overexciting the field, the power factor will lead and the AC stator current again increases.

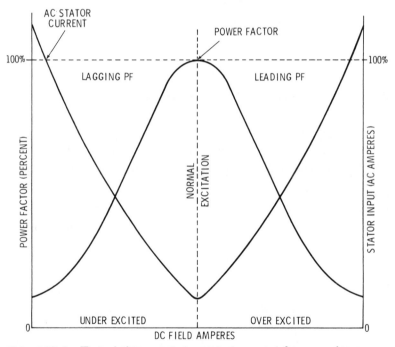

Figure 32-3 Typical characteristic operating curve for a synchronous motor.

Synchronous motors of large size are sometimes run without load on utility lines and the fields are overexcited so as to bring the power factors on the utility lines up to unity (100%). These systems are called *synchronous capacitors*. Industry also uses synchronous motors to correct power factors, as illustrated in Figure 32-4, and the motor usually runs some load, such as a pump.

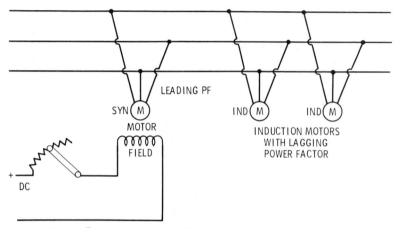

Figure 32-4 Correcting power factor with a synchronous motor.

In DC motors, no power factor is encountered, so they get their regulation slipping more or less and thus reduce or increase the counter emf. This was covered in detail in Chapter 19, "DC Motors."

Regulation

In synchronous motors, the regulation is accomplished by the angle of lag between the alternator and motor poles. Vectors must be used to explain this function. See Figure 32-5,

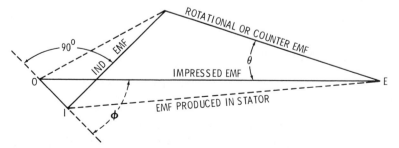

Figure 32-5 Relation of emfs in a synchronous motor.

where

> $O\text{-}E$ = impressed voltage
> $O\text{-}I$ = effective voltage; the component overcoming resistance
> $I\text{-}D$ = inductive drop in the stator windings, which is one of the components of the produced voltage, lagging the current by 90°
> $E\text{-}D$ = rotational or counter emf, the component generated by the rotation of the field or flux cutting the stator conductors
> $E\text{-}I$ = total voltage produced in the stator windings
> ϕ = angle between the impressed voltage and the current
> θ = angle of lag, and angle of time between the moment of impressed emf and maximum counter emf for the same coil group.

If the motor receives additional load, it can't change speed and still run because it will lack torque to carry this additional load. To take care of this problem, the rotor lags back a few degrees in its phase position as shown by $E\text{-}D$ in Figure 32-6. The magnetic circuit becomes less perfect, and the $E\text{-}D$ component of the counter emf will now be generated a little later.

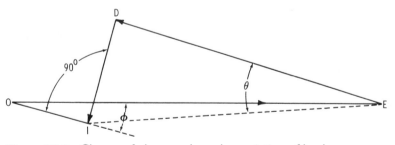

Figure 32-6 Change of phase angle under variation of loads on a synchronous motor.

Since the produced voltage, $E\text{-}I$, depends not only on the value of its two components, $D\text{-}E$ and $D\text{-}I$, but also on their phase relation, the lagging back thus changes the time at which the counter emf will be maximum. The geometric sum of $D\text{-}E$ and $D\text{-}I$ will be less than before, that is, less than $E\text{-}I$ in Figure 32-5. With the reduction of $E\text{-}I$, the effective voltage, $O\text{-}I$, will increase the current. This increased current will have a larger active component and the power factor will improve. Since the current changes, the inductive drop must also change and will, of course, always be 90° from the current. From this it may be readily observed that the motor speed wouldn't have to change, just the angle of lag.

Questions

1. Is there a difference between alternators and synchronous motors? Explain fully.

2. Why is a squirrel cage winding added to a synchronous motor rotor?

3. Draw operating curves for a synchronous motor and explain the operation of the motor.

4. How may lagging power factors of power lines and circuits be corrected by the use of a synchronous motor?

5. Since there can be no slip in a synchronous motor, how does it receive its regulation for variations of loadings? Explain fully.

Chapter 33

Single-Phase Motors

In Chapter 30, covering polyphase motors, the theory of rotating fields was covered. A single-phase motor doesn't in itself create a rotating field as does a polyphase. To ensure that single-phase motors will be self-starting, it is necessary to create a rotating field, at least at start. After they come most of the way up to speed, they will create and maintain their own rotating fields. If a single-phase motor is driven or turned by some means to get it up to approximately 60% to 75% of full speed, it will pick up and run. This arrangement is impractical, so various means have been developed to split a phase, in order to create a rotating field for starting at least.

Phase Splitting

A practical method for illustrating phase splitting will be to use a three-phase motor connected to one phase. If a three-phase motor is running and goes single-phase, it will continue to run but won't pull its rated load and often overheats if the torque requirement is not reduced. On single-phase a polyphase motor will carry 86.6% of full load.

In recent years, phase splitters, made up primarily of capacitors, have been used rather extensively to operate three-phase motors from a single-phase line. They are designed to handle their full rated capacity. These are used often where single-phase is available and three-phase is scheduled in the future. Three-phase motors require considerably less maintenance than single-phase motors.

Figure 33-1 illustrates the simplest means of phase splitting. Winding *A-B* is connected across a single-phase source of power, *D-E*.

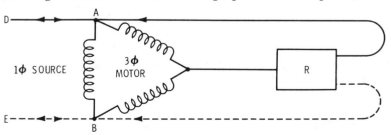

Figure 33-1   Method of starting a three-phase motor on one phase by means of a resistor in shunt with one phase.

Resistor *R*, connected to *C*, may be shunted across windings *A-C* or *B-C*, depending on which direction of rotation is desired. At start the resistor is shunted across one winding, giving that winding a phase current out of phase with the current in winding *A-B*, thus creating a rotating field. As soon as the motor is running, *R* may be disconnected, and the motor will continue to run, making its own rotating field, but will pull only about 86.6% of the full-load rating.

There are other means of phase splitting, such as are shown in Figures 33-2 and 33-3. These are self-explanatory.

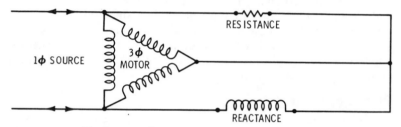

Figure 33-2   Technique of using resistance and inductive reactance for starting a three-phase motor on a single-phase source.

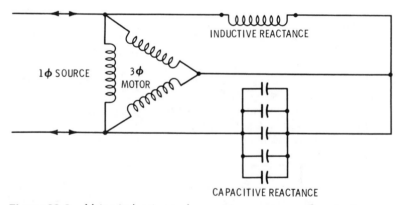

Figure 33-3   Using inductive and capacitive reactances for starting a three-phase motor on one phase.

Shading-Coil Motors

The simplest form of single-phase motor is the shading-coil motor illustrated in Figure 33-4. Many small-fan and clock motors are of this type.

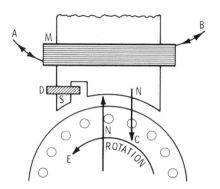

Figure 33-4   Principle of shading-coil self-starting induction motor.

The main coil, *M*, is connected to the AC source, *A-B*. There is one turn, *D*, consisting of a turn of heavy copper conductor short-circuited on itself. The pole piece is notched to accommodate this coil. This gives, as at the instant shown, a large north pole and a small south pole on the same pole piece, thus creating a rotating field. The rotor has a squirrel cage winding. The direction of rotation is always toward the shading coil as shown by *E*.

A four-pole shading-coil motor has four main poles, alternating in polarity, and four shading-coil poles. These are low starting-torque motors and are nonreversible.

Universal Motors

These are single-phase motors designed similarly to DC motors, that is, a wound armature, commutator, and brushes, plus field coils. The field has to be laminated because of the AC and eddy currents. These motors are used for vacuum cleaners, saws, drills, etc. Because of the AC impedance, the field winding has fewer turns than a DC motor of the same rating would have. Universal motors are well adapted for speed control.

Hysteresis Motors

Hysteresis motors are used for clocks and phonographs. The stator is a shading-coil type to make it self-starting. The rotor core is hardened magnetic steel, instead of the usual annealed, low-loss, silicon-steel laminations.

The rotor-core hysteresis loss is greatly magnified, giving the effect of a synchronous motor.

Split-Phase Induction Motor

Split-phase motors are widely used. They are two-winding motors. The running winding is made up of relatively heavy insulated copper conductors, and the starting winding is made up of relatively small-size insulated conductors. The two windings are placed 90 electrical degrees apart. The running winding is at the bottom of the slots, and the starting winding is at the top of the slots.

Figure 33-5 gives the relative positions of the running and starting windings of a four-pole split-phase motor.

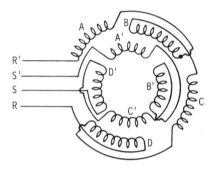

Figure 33-5 Relative positions of running and starting windings of a four-pole split-phase motor.

The running winding is made up of four sets of coils: *A, B, C,* and *D. A* is wound in a reverse direction to *B,* etc.

The starting winding is 90 electrical degrees displaced from the running winding and consists of coils *A′, B′, C′,* and *D′,* with every other coil reversed as was the running winding.

The difference in the conductor sizes of the running and starting windings, as well as the difference in the number of turns in the two respective windings, gives a phase displacement and thus a rotating field.

At start, both the running and starting windings are connected across the line. When the motor gets up to ⅔ to ¾ the running speed, a centrifugal switch opens, cutting the starting winding out of circuit and the motor pulls into its running speed. Figure 33-6

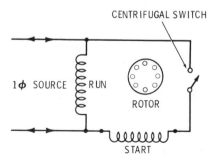

Figure 33-6 Schematic of a split-phase motor.

illustrates a schematic of a split-phase motor, showing the centrifugal switch that cuts the starting winding in or out of the circuit.

To reverse this motor, the starting or running winding is reversed, but not both. Also, the motor must come to a standstill before reversing. See Figure 33-7.

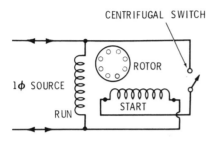

CENTRIFUGAL SWITCH

Figure 33-7 Reversing a split-phase motor.

1φ SOURCE

RUN

ROTOR

START

These motors are also wound for dual voltages, as illustrated in Figure 33-8. Note the starting winding is a 115-volt winding only.

Capacitor-Start Motors

The capacitor-start motor has become an outstanding single-phase motor. It is built very similarly to the split-phase motor, except that the starting winding is heavier. These motors have excellent starting torque. They came into general use with the development of the electrolytic capacitors of large capacitance and of small size. They are practically always dual-voltage motors and can be reversed the same as the split-phase motors. The capacitor is in series with the centrifugal switch, so it is only in circuit during start and for a short period. The number of starts and stops per minute are limited because of the electrolytic capacitor. (See Figure 33-9.) They are reversed in the same manner as the split-phase motor.

Capacitor-Start, Capacitor-Run Motor

This motor has the high starting torque of the capacitor-start motor and is quieter in running than the split-phase motor or the capacitor-start motor, as the running capacitor continues to split a phase and the system is more like a polyphase motor. See Figure 33-10.

There is another version of this motor that uses the capacitor to start and to run, but has no centrifugal switch. The starting torque is low, there is no centrifugal switch, and the capacitor stays in circuit all of the time, so the capacitor can't be of the electrolytic type. See Figure 33-11. These motors are used on fans and oil burners, etc., where low starting torque is required.

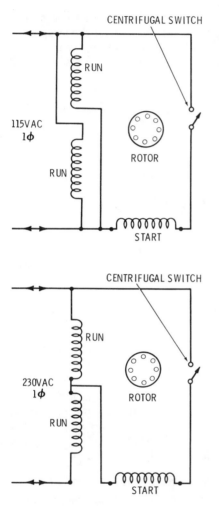

Figure 33-8 Connections for a dual-voltage split-phase motor.

Split-phase and capacitor-type motors are all two-winding motors and are reversed in the same manner as was described previously.

Repulsion Motors

On the repulsion motor the stator is wound with running windings only, and the rotor is an armature (wound) with a commutator. The

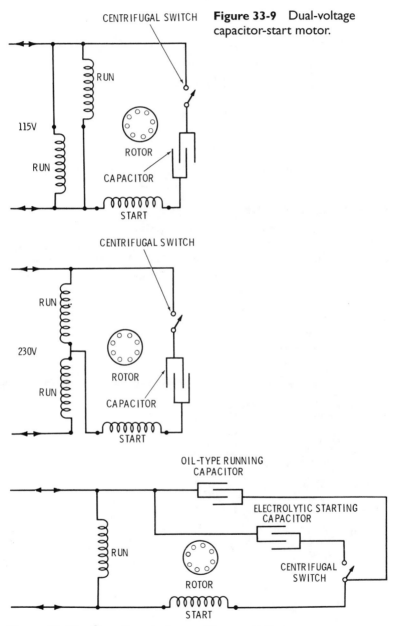

CENTRIFUGAL SWITCH

Figure 33-9 Dual-voltage capacitor-start motor.

RUN

115V

RUN

ROTOR

CAPACITOR

START

CENTRIFUGAL SWITCH

RUN

230V

RUN

ROTOR

CAPACITOR

START

OIL-TYPE RUNNING CAPACITOR

ELECTROLYTIC STARTING CAPACITOR

RUN

ROTOR

CENTRIFUGAL SWITCH

START

Figure 33-10   Capacitor-start, capacitor-run motor.

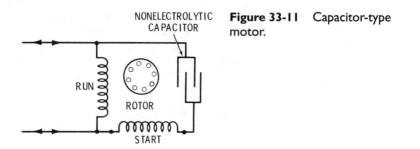

Figure 33-11 Capacitor-type motor.

commutator brushes are short-circuited. The direction of rotation is reversed by brush shifting. See Figure 33-12.

Repulsion-Induction Motors

These motors are very similar to the repulsion motor with shorted brushes. The difference is that as the motor comes up to speed, a centrifugal mechanism moves a short-circuiting necklace out, shorting all of the commutator bars, and it then runs as an induction

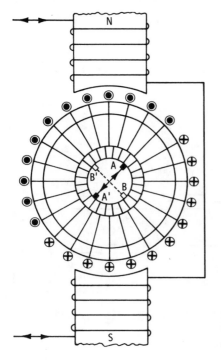

Figure 33-12 Repulsion motor—shifting short-circuited brushes from A-A' to B-B' changes direction of rotation.

motor. The starting torque of both the repulsion motor and the repulsion-induction motor is high. Both of these types of motors have been replaced to a large degree by the capacitor-start motors.

Questions

1. Can a three-phase motor be started on one phase and, if so, how? Explain fully.
2. Describe the shading-coil motor, and explain how it operates.
3. Explain what a universal motor is.
4. Describe a hysteresis motor and how it operates.
5. What is a split-phase motor?
6. Sketch a dual-voltage, split-phase motor.
7. How do you reverse a split-phase motor?
8. Describe dual-voltage, capacitor-start motors.
9. Describe a capacitor-start, capacitor-run motor.
10. Describe a repulsion motor.
11. Describe a repulsion-induction motor.

Chapter 34

Grounding and Ground Testing

Grounding and ground testing seem to mystify many in the electrical industry. This need not be so. Grounding is as good as the precautions that are taken to see that there is a low resistance of the electrical ground circuits to earth.

Grounding Requirements

Article 250 of the *National Electrical Code* (*NEC*) describes what must be done to satisfy the grounding requirements of the *NEC* and is very complete. Those working with grounding and who wish further interpretation may find this in Audel's *Guide to the National Electrical Code,* which is kept current with revisions of the *NEC.*

Section 250.84 of the *NEC* tells us the following: "A single electrode consisting of a rod, pipe, or plate which doesn't have a resistance to ground of 25 ohms or less shall be augmented by one additional electrode of any of the types specified in Section 250.81 or 250.83. Where multiple rod, pipe, or plate electrodes are installed to meet the requirements of this section, they shall be not less than 6 feet (1.83 m) apart."

The purpose of this chapter will be to determine if the requirements of the *NEC* have been met. Actual tests are the only way to determine if the *NEC* requirements have been met. As an inspector of many years in the field, the author feels that the requirements for testing are a necessity to protect people and their properties properly.

There are basically three fundamental reasons for grounding:

1. For protection from overvoltage, should the primary and secondary of the transformer become crossed, or should high-voltage lines cross with low-voltage lines.

2. To dissipate lightning or static charges or any other types of surge voltages.

3. To place non-current-carrying parts of an electrical system at zero potential to ground. Here, when the word "ground" is used, it may mean earth, concrete, walls, floors, piping, etc.

Conduction in Soil

The resistivity of soils varies with four basic conditions: (1) soil composition, (2) moisture content, (3) salts and mineral content, and (4) temperature.

Water is often falsely assumed to be a good conductor. Water is basically an insulator. It becomes a conductor with the addition of impurities such as salts and minerals. As an example, the statement is very often heard, "There is a good ground because the ground rod is in water and gravel." This statement is probably more often false than true, as will be seen from the tables that will be presented as this coverage progresses.

First, in dealing with resistivity of soils, the term ohm-centimeter (ohm-cm) is identified as the resistance of a cube of material (in this case, earth soil), with the cube sides being measured in centimeters.

Table 34-1   Resistivity of Different Soils

Soil	Restivity (ohm-cm)		
	Average	Min.	Max.
Fills-ashes, cinders, brine wastes	2370	590	7000
Clay, shale, gumbo, loam	4060	340	16,300
Same, with varying proportions of sand and gravel	15,800	1020	135,000
Gravel, sand, stones, with little clay or loam	94,000	59,000	458,000

Courtesy Biddle Instruments.

In Table 34-1 there are three columns of resistivity: average, minimum, and maximum. In this table the soils with the least resistance top the list and, finally, at the bottom, there appears gravel, sand, and stones, with little clay or loam. This bears out what was said about sand and gravel making for a poor ground.

Table 34-1 is representative and is taken from *Getting Down to Earth,* a booklet published by Biddle Instruments, 510 Township Line Road, Blue Bell, PA 19422. It is reproduced here courtesy of this company.

Water was also mentioned. Table 34-2 covers resistivity of topsoil and sandy loam. This table was also taken from *Getting Down to Earth.*

Table 34-2   Effect of Moisture Content on Earth Resistivity

Moisture Content Percent by Weight	Resistivity (ohm-cm)	
	Topsoil	Sandy Loam
2.5	1000×10^6	1000×10^6
0	250,000	150,000
5	165,000	43,000
10	53,000	18,500
15	19,000	10,500
20	12,000	6300
25	6400	4200

Courtesy Biddle Instruments.

To further back up the previous statement concerning water and resistivity, Table 34-3 is presented, also taken from *Getting Down to Earth*. In this table, note the reduction of the resistivity (ohmcentimeters) as salt is added.

Table 34-3   Effect of Salt Content on Earth Resistivity*

Added Salt Percent by Weight of Moisture	Resistivity (ohm-cm)
0	10,700
0.1	1800
1.0	460
5	190
10	130
20	100

For sandy loam—moisture content, 15% by weight, temperature 17°C (63°F).
Courtesy James G. Biddle Co.

From these tables, one can readily observe that the combination of sand and gravel is a poor conductor, as is moisture without impurities such as salts or minerals. The point is, how does one know what the resistivity of the soil is, without proper testing? In short, when is a ground a proper ground??

The *NEC* recognizes both exothermic welding and clamps listed for the purpose for connecting grounding electrode conductors to either the copper buried in concrete or the rebar. It is the author's opinion that the exothermic welding (cadwelding) is the better of the two methods; both are usually embedded in concrete, but with cadwelding proper testing will show that the connection is properly made and will stay that way.

It is known that concrete draws moisture and is a conductor of electricity. The Dupont Company has run tests on this, and for those who may wish further information on this subject, it may be found in "The Use of Concrete Enclosed Reinforcing Rods as Grounding Electrodes," a paper published by E. J. Fagan, Member, and R. H. Lee, Senior Member, of E. I. Dupont de Nemours and Co., Wilmington, DA.

There are no tables that will accurately give soil resistivity for the location in which a specific grounding electrode is to be used. Tables are guidelines and the installer must be in a position to run the ground resistance tests. The author, having been responsible for observing and also running many ground resistance tests, has observed that very few electricians have ever been involved in a ground test. Therefore the mechanics of ground resistance testing will be covered in this chapter.

Grounding Electrode

The *NEC* recognizes the use of a copper conductor embedded in the lower portion of building footings as an approved grounding electrode. This is termed the *Uffer Ground* because of the experimentation with favorable results, by Mr. Uffer, in attempting to get a ground of low resistivity.

This is further noted in the *NEC* by permitting rebar in the lower portions of building footings to be used as an approved grounding electrode.

It has been the author's privilege to be closely associated with this type of grounding procedure. In one case, caissons with two 1-inch steel bars extending to the bottom were used as electrodes. For the most part, $2/3$ of the depth was sand and gravel and $1/3$ clay and loam. Tests have run as low as 0.1 ohm per caisson, to an average of possible 2.5 ohms per caisson, the overall test coming out in tenths of ohms. It is easy to see that this is lower than the three ohms referred to in the *NEC* for buried water piping. Figure 34-1 illustrates caisson steel being used as grounding electrodes.

Fall-of-Potential Method

Figure 34-2 illustrates the setup for using a Biddle Ground Megger®. This instrument is marked C_1, P_1 and P_2, C_2, P standing for potential and C for current. This instrument consists of a DC

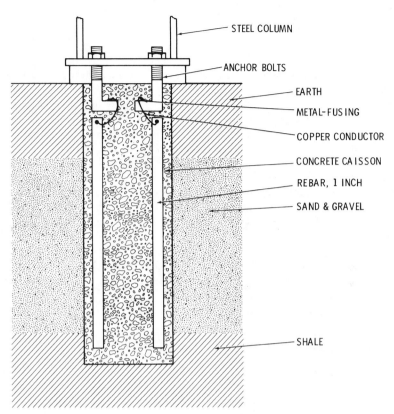

STEEL COLUMN

ANCHOR BOLTS

EARTH

METAL-FUSING

COPPER CONDUCTOR

CONCRETE CAISSON

REBAR, 1 INCH

SAND & GRAVEL

SHALE

Figure 34-1   Caisson rebar grounding to building steel.

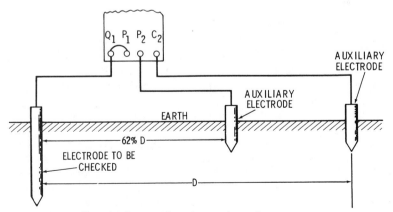

Q_1 P_1 P_2 C_2

AUXILIARY
ELECTRODE

AUXILIARY
ELECTRODE

EARTH

62% D

ELECTRODE TO BE
CHECKED

D

Figure 34-2   Standard setup for electrode resistance test.

generator, hand-cranked, motor-driven, or rectifier-operated, and the meter has a current coil and a potential coil, which together give $R = E/I$, with the reading directly in megohms.

Note that C_1 and P_1 are connected together and in turn connected to the electrode being tested. C_2 is a short electrode driven into the earth at distance D, and P_2 is another short electrode driven into the earth at 62% of distance D from the electrode under test. Thus, drive electrode C_2 at 100 ft from the electrode being tested and P_2 at 62 ft from the electrode being tested.

If P_2 at C_2 are too close together or too close to the electrode being tested, the fields covered by the electrode being tested and of C_2 would overlap as shown in Figure 34-3 and give a false reading.

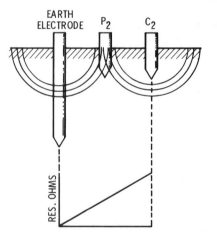

Figure 34-3 Overlapping of test field.

By separating all electrodes as previously described, the results in Figure 34-4 will be obtained. The two fields are separated by sufficient distance so they don't affect one another. In looking at the graph, the reason for the 62% distance for electrode P_2 may be seen: It is the knee X of the curve, which is the straightest portion of the curve.

There are times when one must take ground resistance measurements from inside a building and the C_1P_1 lead is not long enough to reach the point to be tested. In this case, take a conductor, say, No. 12 copper, x number of feet in length to reach the point that is being tested. The resistance of this extra conductor must be determined in order that it may be subtracted from the test reading on

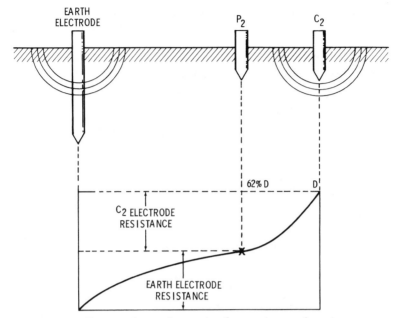

Figure 34-4 Proper electrode spacing for testing earth resistance.

the meter, to give the actual resistance to earth of the point being tested. To arrive at the resistance of this extra conductor, stretch it out in a straight loop, as illustrated in Figure 34-5, connecting one end to C_1P_1 and the other end to P_2C_2, and take the meter reading, which will be the resistance of this auxiliary conductor.

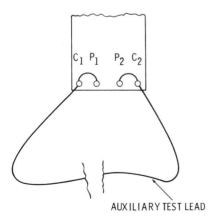

AUXILIARY TEST LEAD

Figure 34-5 Method of testing auxiliary length of test lead.

Now, if the ground electrode under test doesn't meet the required resistance, the following methods may be used to lower the resistance:

1. Parallel additional grounding electrodes.

2. Use longer electrodes.

3. Use chemical additives in the soil near the electrode.

Two rods paralleled 5 ft apart reduce the original resistance to about 65%. Three rods paralleled 5 ft apart will reduce the original resistance to about 42%, and four rods paralleled 5 ft apart will reduce the original resistance to approximately 30%.

The results of additional depths of driving the electrode may be seen in Figure 34-6. It becomes evident from the graph that any increase in length beyond 10 ft doesn't have much effect in reducing the resistance.

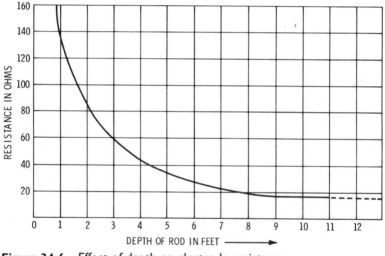

Figure 34-6   Effect of depth on electrode resistance.

There are methods of reducing the resistance of the earth around a made electrode. They are not recommended as a general practice, as they are not permanent, and it is far better to go the permanent route in securing good grounds, unless the earth is under a controlled system of periodic inspections. Figure 34-7 shows a typical layout.

Of the three methods of reducing the grounding electrode resistance discussed above, the driving of additional rods is preferred.

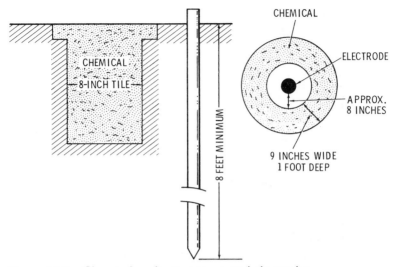

Figure 34-7 Chemical application to ground electrodes.

The *NEC* states that connections to encased rods or rebar shall be made by metal fusing if the connection is encased. Another common expression to indicate metal fusing is *exothermic welding.* In this, metal fusing the iron and copper conductor shall be clean and dry to ensure proper results.

To check a metal-fused connection, the following proves very satisfactory: A source of DC, such as from a DC welding machine, is applied across the weld. See Figure 34-8. The amount of amperage used depends upon the ampacity of the grounding conductor. Satisfactory results can be obtained by using 100 amperes DC with 4/0 copper conductors. The weld resistance shouldn't exceed 0.01 ohm, and good welds will be much less than this. Take the ampere reading and divide it into the voltage drop across the weld and this will give the resistance reading in ohms.

The above method of reading resistance on connections is very satisfactory, but it is cumbersome and time-consuming.

The same results may be obtained by means of a *Ducter*®, which is a registered trademark of Biddle Instruments. This also uses the fall-of-potential method of testing, but instead of using separate voltmeter and ammeter, the results are indicated on a meter directly, by a cross-coil true ohmmeter.

The instrument consists of two coils mounted rigidly together on a common axis in the field of a permanent magnet. The source of

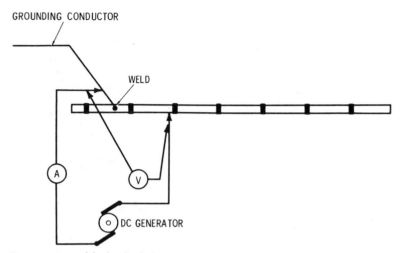

Figure 34-8 Method of checking resistance of metal fusing.

current and voltage may be batteries, chargeable or not, or a rectifier plugged into 120 volts AC.

The Ducter® operates independently of voltage, as the voltage and current vary in direct proportion; thus, no rheostats or other balancing devices are required.

Ducters® have 0- to 10,000- to 1,000,000-microhm readings and operate at 1 to 100 amperes, depending upon the model picked and the range setting. The resistances of the leads supplied are compensated for in the reading obtained, and so don't enter into the readings.

Since the writing of the first edition, Biddle Instruments has come out with a very compact and light low-resistance ohmmeter, DLRO®, which is self-contained and will read down to a half-millionth of an ohm. Biddle Instruments sent me one to evaluate. I took readings previously done with a Ducter® and compared them with the readings on the new instrument, and the results were the same on cadwelds.

System Grounding

The most common industrial distribution systems, as far as grounding is concerned, are

1. Ungrounded (see Figure 34-9)
2. Resistance Grounded (see Figure 34-10)
3. Reactance Grounded (see Figure 34-11)
4. Solid Grounded (see Figure 34-12)

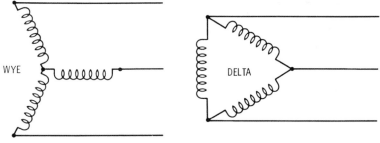

Figure 34-9 Ungrounded systems.

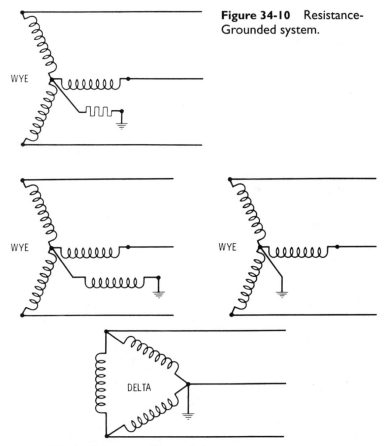

Figure 34-10 Resistance-Grounded system.

Figure 34-11 Reactance-Grounded system.

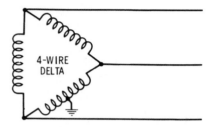

Figure 34-12   Solid-Grounded systems.

The ungrounded system is actually high-reactance capacitance grounded as a result of the capacitive coupling to ground of every energized cable, conductor, bus bar, or machine coil. A single-phase ground won't trip an overcurrent device, but the other two phases are subjected to about 73% overvoltage on a sustained basis and this system may be subjected to voltage as high as five times normal with an arcing fault.

In resistance grounding, the circuit acts more as a resistance than a capacitor and doesn't subject the other phases to overvoltage with a ground fault on one phase. The current, when a ground fault occurs on just one phase, won't trip the circuit. Two phase grounds would cause circuit interruption. This type of system effectively bleeds off disturbing influences. These may be surges by lightning, switching, ferroresonance, etc.

Reactance-grounded circuits are not ordinarily encountered in industrial power systems. Transitory overvoltage will occur on repetitive restriking in an arc on a ground fault.

The solid-grounded system is used extensively in systems of 600 volts or less. It is very effective in giving the greatest control of overvoltages.

Early in this chapter soil resistivity was covered and an explanation given of ohm-cm. It might be good to go further into soil resistivity, as it has a very valuable use.

In the actual testing of the resistance of a grounding electrode to meet the 25-ohm maximum requirement of the *Code*, a ground resistance tester is all that is needed. However, with grounding electrode resistance being more important, especially with the growing use of computers, it may be appropriate to check soil resistivity to locate the most applicable and satisfactory point where the ground electrode will have the lowest resistance.

Another good reason for checking earth resistivity is to find where cathodic protection is required for protection of underground metallic piping, such as water, sewer, oil, gas, and gasoline piping systems, against electrolysis. Other uses of nonelectrical nature may be locating old land fills, ores, and water-bearing ground.

A case in point: The author was hired to check earth resistivity for a computer building to check stack emissions from a power plant, so that the proper grounding system could be developed. In all cases of checking soil resistivity the method is similar, so the overall procedure will be given. For additional information Biddle Instruments may be contacted for the booklet, *Getting Down to Earth.*

Instead of using the three-lead method for checking grounding electrode resistance, you use the four-lead method as shown in Figure 34-13, using terminals C_1, P_1, P_2, and C_2.

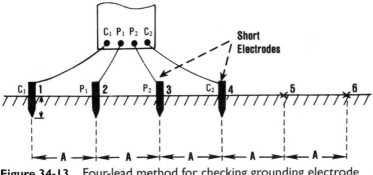

Figure 34-13 Four-lead method for checking grounding electrode resistance.

Four short electrodes are used and the distance B to which the electrodes are drawn is about 6 inches.

The four short electrodes are spaced at a distance at which you wish to check the soil resistivity. It is recommended that a straight line be established for the test to be made on. Use a cord to the electrodes, so the distance is maintained to the depth the earth resistivity is to be checked. Thus, if you wish to check to the depth of 10 ft, the cord maintains the 10-ft distance between the electrodes as they are.

Connect terminal C_1 to 1, terminal P_1 to 2, terminal P_2 to 3, and terminal C_2 to 4. Secure the Megger® reading and write it down as "Test No. 1."

The next move is to drive rod 1 at point 5. Reconnect the leads as follows: terminal C_1 to 2, terminal P_1 to 3, terminal P_2 to 4, and terminal C_2 to 5. Continue this procedure for whatever distance you wish to cover on the testing, recording the Megger® readings as with Test No. 1. Keep the readings in the order that you have taken them.

As an example of this coverage, we will assume the electrodes are spaced 10 ft apart and the reading on Test No. 1 was 70.

Dr. Frank Werner, of the U.S. Bureau of Standards, developed this test in 1915. He showed that if the electrode depth was small, the formula $p = 2AR$ would give the ohm-cm earth resistivity, where

p = ohm-cm

A = distance between electrodes in cm

R = Megger® reading in ohms

The distance A between electrodes for this example was 10 ft. This must be converted to cm (1 inch = 2.54 cm).

120 in. (10 ft × 12 in. per ft) × 2.54 = 304.8 cm

= 3.1417 so:

= 2 × 3.1416 × 304.8 × 70 = 134,058.04 ohm-cm on

Test No. 1.

A graph may be drawn showing the ohm-cm of each setting, which will graphically illustrate the variations of the soil resistivity.

A larger area may be covered by running a line parallel to the line just covered, separating the parallel lines by the distance used between the test electrodes. From this test, a larger area is covered and a more appropriate location for the grounding electrode will be shown. After all, we are looking for the lowest ohm-cm earth resistivity reading for placing the grounding electrode(s).

Questions

1. What are the three fundamental reasons for grounding?
2. Give four items that affect soil resistivity.
3. What is an ohm-centimeter?
4. What is the Uffer ground?
5. Explain the drop-of-potential ground-testing method.
6. Name three methods of improving grounds.
7. Explain metal-fusing resistance testing, of which there are two good methods. Explain both.
8. Give the four most used industrial grounding methods and explain good and bad aspects of each one.

Chapter 35

Rectifiers

A *rectifier* is a device for converting AC into DC. The process of this conversion is termed *rectification.*

One of the earliest observations of this is called the Edison effect. Edison found that when a cathode (hot filament) of an electronic tube is heated and a positive potential applied to the anode of this electronic tube, a current would flow between the cathode and the anode. Thompson discovered, in addition, that electrons flowed from the cathode to the anode. Thus this was one of the first rectifiers to change AC to DC. See Figure 35-1.

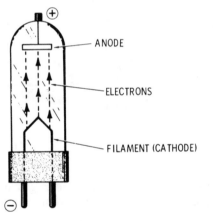

Figure 35-1 Electron tube showing Edison effect.

ANODE

ELECTRONS

FILAMENT (CATHODE)

Mercury-Arc Rectifier

The mercury-arc rectifier was one of the first high-amperage rectifiers. It is also an electronic device consisting of a glass enclosure with a vacuum and a pool of mercury in the bottom, which serves as the cathode, and one or more electrodes made of graphite, which are the anode(s). See Figure 35-2.

To start the rectifier, the glass enclosure is tipped to the right, causing the mercury to flow to C_1, and when the enclosure is straightened up, an arc results. The mercury vapor contacts anodes A_1 and A_2, continuing the flow of electrons, and current flows from the reactor X_L to the battery, B, to charge the battery.

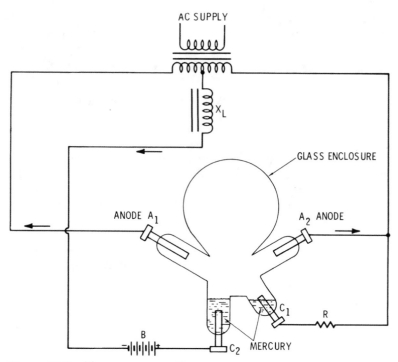

Figure 35-2   Mercury-arc rectifier.

The tungar rectifier is a glass enclosure filled with an inert gas at low pressure, a graphite anode, and a tungsten filament that serves as the cathode. See Figure 35-3.

Dry-Contact Rectifiers

Copper-oxide and selenium rectifiers are called *dry-contact* or *dry-disc* rectifiers. They both use the same principle, that various materials will pass current in one direction but not in the other direction. See Figure 35-4, which illustrates a copper-oxide rectifier.

Figure 35-5 shows a selenium rectifier. It is, as will be noticed, very similar to the copper-oxide rectifier, except for the materials used. The discs of both of these rectifiers may be stacked in series to handle higher voltages, or in parallel to handle larger currents.

Figure 35-6 illustrates the dry-contact rectifier symbol and the direction of the electron flow.

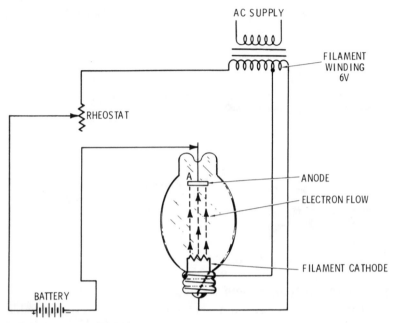

Figure 35-3   Tungar rectifier.

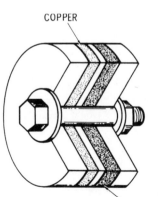

COPPER-OXIDE
COATING

Figure 35-4   Copper-oxide
rectifier.

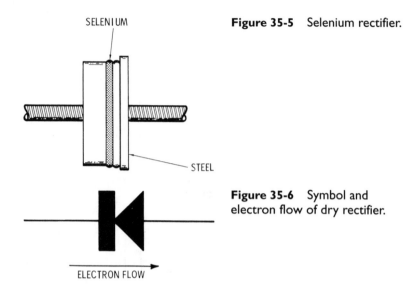

Figure 35-5 Selenium rectifier.

Figure 35-6 Symbol and electron flow of dry rectifier.

Solid-State Rectifiers

Solid-state rectifiers have come into extensive use. They are reliable and may be used for small or large currents. Figure 35-7 illustrates a typical silicon solid-state rectifier.

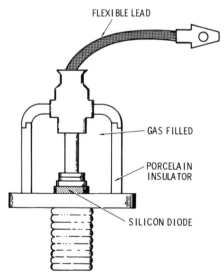

Figure 35-7 Typical silicon solid-state rectifier.

Rectifier Connections

Connections to rectifiers may be made in a number of ways. Figure 35-8 illustrates three ways to connect rectifiers into a circuit and the accompanying waveforms.

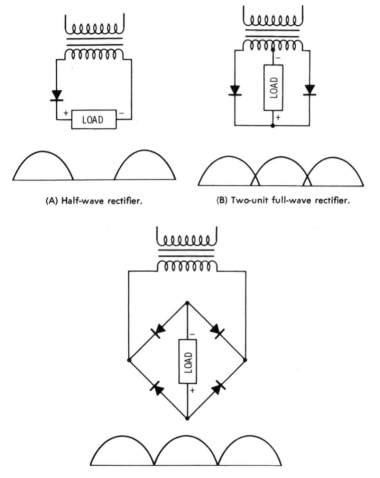

(A) Half-wave rectifier.　　　　(B) Two-unit full-wave rectifier.

(C) Four-unit full-wave rectifier.

Figure 35-8　Rectifier circuit connections.

In order to straighten out the pulsating DC, three-phase power is converted to six-phase power by means of a transformer. Rectification takes place in a six-anode mercury-arc rectifier. This smoothes out the resultant DC wave considerably. See Figure 35-9. This type of rectifier is usually large and the enclosure is steel, with a vacuum pump to keep the enclosure under a vacuum.

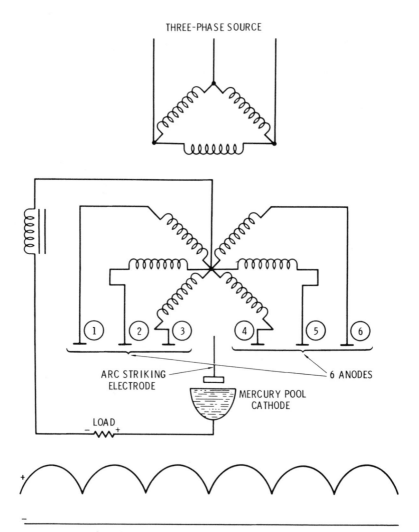

Figure 35-9 Six-phase mercury-arc rectifier.

Questions

1. What is a rectifier?
2. Illustrate and describe the Edison effect.
3. Illustrate and explain the operation of a mercury-arc rectifier.
4. Describe the tungar rectifier.
5. Explain the operation of both the copper-oxide and selenium rectifiers.
6. Draw the symbol and show the electronic flow of dry rectifiers.
7. Draw a diagram of connections for a half-wave rectifier. Also show the typical waves produced.
8. Draw diagrams of both-wave and four-wave rectifiers.
9. Sketch a six-phase mercury-arc rectifier and the output wave form.

Chapter 36

Number of Bends in Conduit

The *NEC* specifies that there shall be no more than 360° total bends in runs of conduit, and this includes all types of conduits. This, of course, is intended to be the maximum degrees allowable, but doesn't mean that you should, or always can, safely go to the maximum of 360° in a run of conduit. There are times when 360° total in bends might be too much and cause damage to the cable insulation or stretching of the conductors.

The coverage in this article is for cables, and the term *cables* is a general term and applies to larger-size conductors. Cables may be

1. A stranded conductor (single-conductor cable)
2. A combination of conductors insulated from each other (multiple-conductor cable)

This is pointed out because small-size conductors don't have a maximum breaking strength equal to what will be used in this article. The formulas included might be adaptable to smaller conductors by taking into consideration the breaking point of the conductors and fitting them into the formulas given here.

As will be shown, the radius of bends or elbows will have much to do with pulling tension. Also, the end from which you pull will affect the tension required in the pull.

Examples and formulas for figuring the tension required to make various pulls will be given, so that inspectors may check the design to see if the actual pulling tension required exceeds the maximum tension permissible and aid in establishing points where pull boxes should be installed.

In this discussion, you will see the advantages of using pulling eyes instead of basket grips (socks) on cables. The pulling eye pulls on the conductor itself and the basket grip pulls on the insulation. It shouldn't be presumed that the basket grip should never be used, especially on short runs. The information given here will permit you to make the proper decision as to which type to use.

The formulas and table used here have been taken from the Okinite company's *Engineering Data for Electrical Cables*.

The maximum stresses must not be exceeded when pulling a cable. The following formulas and information will be given and referred to in the calculations by the numbers assigned to them.

1. The maximum stress shan't exceed 0.008 times the *CM* when pulling with a pulling eye attached to copper or aluminum conductors, $T_m = 0.008 \times n \times CM$ where

    T_m = maximum tension in pounds

    n = number of conductors in cable, (if single-conductor cables, use the number of single-conductor cables)

    CM = Circular mil area of each conductor

2. The maximum stress for lead cables shan't exceed 1500 lb/sq. in of lead sheath area when pulled with a basket grip.

    $$T_m = 4712t \, (D - t)$$

    where

    t = lead sheath thickness, in inches

    D = outside diameter of cable, in inches

3. The maximum stress shan't exceed 1000 lb for non-lead cables when pulled with a basket grip. (However, maximum stress calculated from item (1) can't be exceeded.)

4. The maximum stress at a bend shan't exceed 100 times the radius of curvature of the duct expressed in feet (but maximum stress calculated for item (1), (2), or (3) can't be exceeded).

5. For a straight section, the pulling tension is equal to the length of the duct run in feet, multiplied by the weight per foot of the cable(s) and the coefficient of fiction, which for well-constricted ducts may be taken as 0.5 (for cable with insulation that tends to stick a little, the coefficient of 0.75 should be used).

    $T = L \times w \times f$
    T = total pulling stress in lb
    L = length of straight duct in ft
    w = weight of cable(s) in lb per ft
    f = coefficient of friction, either 0.5 or 0.75

6. For ducts having curved sections, the following formula applies:

    $$T = T_2 + T_1 e^{fa}$$

where

T_2 = tension for straight section at pulling end
T_1 = tension for straight section at feeding end
f = coefficient of friction = 0.5 or 0.75
e = naperian logarithm base = 2.718
a = angle of bend in radians (1 radian = 57.3°)

$$e^{fa} = \log_{10}^{-1}\frac{fa}{2.303}$$

The number whose logarithm to the base 10 is $fa/2.303$

$90° = 1.571$ radians
$45° = 0.7854$ radians
$30° = 0.5236$ radians
$15° = 0.2618$ radians

A table is given below for the coefficients of friction for various angles; it is derived from the above formula. To illustrate how the coefficients of friction were derived in the table, the following explanation may be helpful.

a = angle of bend in radians

$180° = 3.1416$ radians   [this is also π (pi)]

Example

$$\text{Radians in } 90° = \frac{90}{180} \times 3.1416 = 1.5708 \text{ radians}$$

$$\text{Radians in } 15° = \frac{15}{180} \times 3.1416 = 0.2618 \text{ radians}$$

For the coefficient of friction for 90°, using $f = 0.5$:

$$\frac{1.5708 \times 0.5}{2.303} = 0.34103$$

which is the logarithm of a number to the base 10. This number may be found in a common logarithm table or with a calculator, and the number will be 2.19, the same as shown in Table 36-1:

Table 36-1   Pulling Tension Coefficients

Angle of Bend in Degrees	Coefficient of Friction	
	$f = 0.50$*	$f = 0.75$
15	1.14	1.22
30	1.30	1.48
45	1.48	1.81
60	1.68	2.20
90	2.19	3.24

*Generally used.

Bending Example No. 1

We will use six 500-MCM THW cables in one conduit or duct.

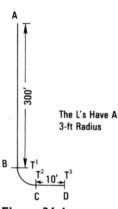

The L's Have A 3-ft Radius

Figure 36-1
Bending example No. 1.

From (1), $T_m = 0.008 \times n \times CM = 0.008 \times 6 \times 500,000 = 24,000$ lb maximum stress with a pulling eye used.

From (3), $T_m = 1000 \times 6 = 6000$ lb maximum stress with a basket grip.

In this example we will use the six 500-MCM THW cables that, according to the *NEC*, will go in a 4″ conduit, and Table 346.10 of the *NEC* tells us the radius of a standard bend is 2 ft, but we will use a radius of 3 ft.

See Figure 36-1. In this case, we will pull at D and start the cable at A; 500-MCM THW copper cable weighs 1.77 lb per ft, so six cables will weight $6 \times 1.77 = 10.62$ lb per ft.

Stress (5) at $T_1 = L \times w \times f = 300 \times 10.62 \times 0.5 = 1593$ lb

Stress (6) at $T_2 = T_1 \times 2.19 = 1593 \times 2.19 = 3489$ lb

Item (4) says the maximum stress at a bend shan't exceed 100 times the radius of the curvature of the duct, so 3 ft $\times$ 100 = 300 lb maximum stress at the 90° elbow. As it is, the stress will be 3489 lb from A to C, which is in excess of the allowable 300-lb maximum stress allowed for a 90° elbow with a 3-ft radius.

Now let's pull at A and start cable at D.

Stress at $T_2 = L \times w \times f = 10 \times 10.62 \times 0.5 = 53.1$ lb.

At B, stress $T_1 = 53.1 \times 2.19 = 116.29$ lb, and this meets the 300-lb maximum stress around the 3-ft-radius elbow, so we can continue to pull at A.

Stress at $A = 116.29 + (L \times w \times f) = 116.29 + (300 \times 10.62 \times 0.5) = 116.29 + 1593 = 1709.29$ lb.

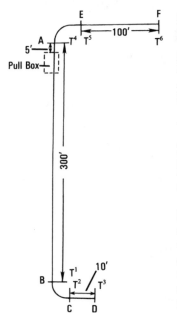

Thus, pulling from D to A, either a pulling eye or a basket grip could be used.

Bending Example No. 2

Next, add a 3-ft radius 90° elbow at A and a 100 ft of conduit from E to F. See Figure 36-2.

In the last example, we found 1709.29-lb pulling stress from D to A. This already exceeds the permissible stress for the elbow AE, so a pull box is required near A, between A and B, say, 5 ft away from A.

Then, by pulling from D to the pull box no tensions are exceeded and the cable being pulled around elbow BC won't exceed the permissible pressure against the conduit; thus, the conductors may be pulled out at the pull box and another pull started from the box to F, or they may be cut and spliced in the pull box with approved splicers. Either way, the stress from the pull box to F will be the same.

Figure 36-2 Bending example No. 2.

$T_4 = 5 \times 10.62 \times 0.5 = 26.5$ lb stress

$T_5 = T_4 + (T_4 \times 2.19) = 26.5 + (26.5 \times 2.19) = 84.5$ lb

Stress at

$F = 84.5 + (L \times w \times f) = 84.5 + (100 \times 10.62 \times 0.5) = 615.5$ lb

From the above examples, you may plainly see that sometimes one 90° elbow is all that can safely be installed in a run of conduit.

The next example will show how bends in lesser degrees and longer radius affect the pulling tension.

Bending Example No. 3

We will use the six 500-MCM THW conductors as before. See Figure 36-3. The pulling will be at D and the conductors inserted at A. The maximum allowable stress at the bend BC is $100 \times 80 = 8000$ lb stress.

Stress at $B = T_1 = L \times w \times f = 300 \times 10.62 \times 0.5 = 15.93$ lb

Stress at $C = T_2 = T_1 \times e^{fa} = 1593 \times 1.14 = 1816.02$ lb

Stress at $D = T_2 = (L_2 \times w \times f)$
$= 1816.02 + (100 \times 10.62 \times 0.5)$
$= 1816.02 + 531 = 2347.02$ lb

Figure 36-3   Bending example No. 3.

Permissible maximum tension

Using pulling eye $(1) = 0.008 \times 6 \times 500{,}000 = 24{,}000$ lb

Permissible maximum tension

Using basket grip $= 1000 \times 6 = 6000$ lb

Permissible maximum tension

At point of curve BC $(4) = 100 \times 80 = 8000$ lb

Thus, a pulling eye or a basket grip may be used. Now, if we pull at A and start the conductors at D,

$T_2 = L \times w \times f = 100 \times 10.62 \times 0.5 = 531$ lb stress

Stress at B $(T_1) = T_2 \times e^{fa} = 531 \times 1.14 = 605.34$ lb

Stress at $A = T_1 + (L \times w \times f)$
$= 605.34 + (300 \times 10.62 \times 0.5) = 2198.34$ lb

From these examples it may be seen that there is less tension (2198.34 lb) required by pulling at A than there is by pulling at D (2347.02 lb).

Chapter 37

Torque Test for Bolts and Screws

I have personally found very little information on torquing values available, so it might be appropriate to insert some torquing values in this book (Tables 37-1–37-3). Many breakdowns and possible fires might result from not adhering to proper torquing values, so the following tables are presented as a guideline for tightening connections. It might also be mentioned that dies on compression tools do wear, and to avoid breakdowns, Biddle Instrument's Ductor® can prevent this problem, as it will read down to a half-millionth of an ohm. This instrument has been invaluable to me on jobs in the past.

Loose connections can be a hazard, causing breakdowns and possibly fires.

If the authorities having jurisdiction desire this, they may require torquing tests during inspections.

Table 37-1   Tightening Torque in Pound-Feet—Screw Fit

Wire Size	Driver	Bolt	Other
18–16			4.2
14–8	1.67	6.25	6.125
6–4	3	12.5	8.0
3–1	3.2	21.00	10.40
0–2/0	4.22	29	12.5
AWG 200MCM	—	37.5	17.0
250–300	—	50.0	21.0
400	—	62.5	21.0
500	—	62.5	25.0
600–750	—	75.0	25.0
800–1000	—	83.25	33.0
1250–2000	—	83.26	42.0

Table 37-2   Screws

Screw Size in Inches Across Hex Flats	Torque in Pound-Feet
$1/8$	4.2
$5/34$	8.3
$3/16$	15
$7/32$	23.25
$1/4$	42

Table 37-3   Bolts

Standard, Unlubricated			
Size	Duronze	Steel	Aluminum
$3/8$	20	15	16
$1/2$	40	25	35
$5/8$	70	50	50
$3/4$	100	90	70
Lubricated			
$3/8$	15	10	13
$1/2$	30	20	25
$5/8$	50	40	40
$3/4$	85	70	60

Appendix

Table A-1   Trigonometric Functions (Natural)

Angle	Sine	Cosine	Tangent	Angle	Sine	Cosine	Tangent
0°	0.000	1.000	0.000°	31°	.515	.857	.601
1°	.018	1.000	.018	32°	.530	.848	.625
2°	.035	0.999	.035	33°	.545	.839	.649
3°	.052	.999	.052	34°	.559	.829	.675
4°	.070	.998	.070	35°	.574	.819	.700
5°	.087	.996	.088	36°	.588	.809	.727
6°	.105	.995	.105	37°	.602	.799	.754
7°	.122	.993	.123	38°	.616	.788	.781
8°	.139	.990	.141	39°	.629	.777	.810
9°	.156	.988	.158	40°	.643	.766	.839
10°	.174	.985	.176	41°	.656	.755	.869
11°	.191	.982	.194	42°	.669	.743	.900
12°	.208	.978	.213	43°	.682	.731	.933
13°	.225	.974	.231	44°	.695	.719	.966
14°	.242	.970	.249	45°	.707	.707	1.000
15°	.259	.966	.268	46°	.719	.695	1.036
16°	.276	.961	.287	47°	.731	.682	1.072
17°	.292	.956	.306	48°	.743	.669	1.111
18°	.309	.951	.325	49°	.755	.656	1.150
19°	.326	.946	.344	50°	.766	.643	1.192
20°	.342	.940	.364	51°	.777	.629	1.235
21°	.358	.934	.384	52°	.788	.616	1.280
22°	.375	.927	.404	53°	.799	.602	1.327
23°	.391	.921	.425	54°	.809	.588	1.376
24°	.407	.914	.445	55°	.819	.574	1.428
25°	.423	.906	.466	56°	.829	.559	1.483
26°	.438	.899	.488	57°	.839	.545	1.540
27°	.454	.891	.510	58°	.848	.530	1.600
28°	.470	.883	.532	59°	.857	.515	1.664
29°	.485	.875	.554	60°	.866	.500	1.732
30°	.500	.866	.577	61°	.875	.485	1.804

(continued)

Table A-1 (continued)

Angle	Sine	Cosine	Tangent	Angle	Sine	Cosine	Tangent
62°	.883	.470	1.881	77°	.974	.225	4.331
63°	.891	.454	1.963	78°	.978	.208	4.705
64°	.899	.438	2.050	79°	.982	.191	5.145
65°	.906	.423	2.145	80°	.985	.174	5.671
66°	.914	.407	2.246	81°	.988	.156	6.314
67°	.921	.391	2.356	82°	.990	.139	7.115
68°	.927	.375	2.475	83°	.993	.122	8.144
69°	.934	.358	2.605	84°	.995	.105	9.514
70°	.940	.342	2.747	85°	.996	.087	11.43
71°	.946	.326	2.904	86°	.998	.070	14.30
72°	.951	.309	3.078	87°	.999	.052	19.08
73°	.956	.292	3.271	88°	.999	.035	28.64
74°	.961	.276	3.487	89°	1.000	.018	57.29
75°	.966	.259	3.732	90°	1.000	.000	Infinity
76°	.970	.242	4.011				

Table A-2   Ratings of Conductors and Tables to Determine Volt Loss*

How to Figure Volt Loss	*How to Select Size of Wire*
Multiply distance (length in feet of one wire) by the current (expressed in amperes)	Multiply distance (length in feet of one wire) by the current (expressed in amperes).
By the figure shown in table for the kind of current and the size of the wire to be used.	Divide that figure into the permissible volt loss multiplied by 1,000,000.
Then put a decimal point in front of the last 6 figures—you have the volt loss to be expected on that circuit.	Look under the column applying to the type of current and power factor for the figure nearest, but not above your result. You have the size of wire needed.
Example—No. 6 copper wire in 180 feet of iron conduit—3-phase, 40-amp load at 80% power factor.	**Example**—Copper wire in 180 feet of iron conduit—3-phase, 40-amp load at 80% power factor—volt loss from local code 5.5.
Multiply feet by amperes: 180 × 40 = 7200.	Multiply feet in amperes: 180 × 40 = 7200.
Multiply this number by number from table for No. 6 wire 3-phase at 80% power factor: 7200 × 735 = 5,292,000.	Divide permissible volt loss multiplied by 1,000,000 by this number: 5.5 × 1,000,000/7200 = 764.
Place decimal point 6 places to left. This gives volt loss to be expected: 5.292 volts.	Select number from table, 3-phase at 80% power factor that is nearest, but not greater than, 764. This number is 735, which indicates the size of wire needed: No. 6.
(For a 240-volt circuit, the percent voltage drop is 5.292/240 × 100, or 2.21%.)	

*With higher ratings on new insulations, it is extremely important to bear volt loss in mind; otherwise, some very unsatisfactory experiences are likely to be encountered.

These tables take into consideration reactance on AC circuits as well as resistance of the wire.

Remember on short runs to check to see that the size and the type of wire indicated have sufficient ampere capacity.

Table A-3 Copper Conductors

Copper Conductors in Iron Conduct

Wire Size	Type T, TW (60°C Wire)	Type RH, THWN, RHW, THW (75°C Wire)	Type RHH, THHN, XHHW (90°C Wire)	Direct Current	Three-Phase—60 Hz, Lagging Power Factor					Single-Phase—60 Hz, Lagging Power Factor				
					100%	90%	80%	70%	60%	100%	90%	80%	70%	60%
14	15	15	15	6100	5280	4800	4300	3780	3260	6100	5551	4964	4370	3772
12	20	20	20	3828	3320	3030	2720	2400	2080	3828	3502	3138	2773	2404
10	30	30	30	2404	2080	1921	1733	1540	1340	2404	2221	2003	1779	1547
8	40	45	50	1520	1316	1234	1120	1000	880	1520	1426	1295	1159	1017
6	55	65	70	970	840	802	735	665	590	970	926	850	769	682
4	70	85	90	614	531	530	487	445	400	614	613	562	514	462
3	80	100	105	484	420	425	398	368	334	484	491	460	425	385
2	95	115	120	382	331	339	322	300	274	382	392	372	346	317
1	110	130	140	306	265	280	270	254	236	306	323	312	294	273
0	125	150	155	241	208	229	224	214	202	241	265	259	247	233
00	145	175	185	192	166	190	188	181	173	192	219	217	209	199
000	165	200	210	152	132	157	158	155	150	152	181	183	179	173
0000	195	230	235	121	105	131	135	134	132	121	151	156	155	152
250M	215	255	270	102	89	118	123	125	123	103	136	142	144	142
300M	240	285	300	85	74	104	111	112	113	86	120	128	130	131
350M	260	310	325	73	63	94	101	105	106	73	108	117	121	122
400M	280	335	360	64	55	87	95	98	100	64	100	110	113	116
500M	320	380	405	51	45	76	85	90	92	52	88	98	104	106
600M	355	420	455	43	38	69	79	85	87	44	80	91	98	101
700M	385	460	490	36	33	64	74	80	84	38	74	86	92	97

750M	400	475	500	34	31	62	72	79	82	36	72	83	91	95
800M	410	490	515	32	29	61	71	76	81	33	70	82	88	93
900M	435	520	555	28	26	57	68	74	78	30	66	78	85	90
1000M	455	545	585	26	23	55	66	72	76	27	63	76	83	88

Copper Conductors in Nonmagnetic Conduit

				6100	5280	4790	4280	3760	3240	6100	5530	4936	4336	3734
14	15	15	15	6100	5280	4790	4280	3760	3240	6100	5530	4936	4336	3734
12	20	20	20	3828	3320	3020	2700	2380	2055	3828	3483	3112	2742	2369
10	30	30	30	2404	2080	1910	1713	1513	1311	2404	2202	1978	1748	1512
8	40	45	50	1520	1316	1220	1100	976	851	1520	1406	1268	1128	982
6	55	65	70	970	840	787	715	641	562	970	908	825	740	648
4	70	85	90	614	531	517	466	422	374	614	596	538	486	431
3	80	100	105	484	420	410	379	344	308	484	474	438	397	355
2	95	115	120	382	331	326	303	278	250	382	376	350	321	288
1	110	130	140	306	265	266	251	232	211	306	307	289	267	243
0	125	150	155	241	208	216	206	192	176	241	249	237	221	203
00	145	175	185	192	166	176	170	160	148	192	203	196	184	171
000	165	200	210	152	132	145	141	134	126	152	167	163	155	145
0000	195	230	235	121	105	119	118	114	108	121	137	136	131	125
250M	215	255	270	102	89	105	106	104	100	103	121	122	120	115
300M	240	285	300	85	74	92	95	93	91	86	106	109	107	105
350M	260	310	325	73	63	82	85	84	83	73	94	98	97	96
400M	280	335	360	64	55	75	78	79	78	64	86	90	91	90
500M	320	380	405	51	45	64	69	71	70	52	74	80	82	81
600M	355	420	455	43	38	57	63	66	66	44	66	73	76	76
700M	385	460	490	36	33	53	58	61	63	38	61	67	70	73
750M	400	475	500	34	31	51	56	60	61	36	59	65	69	70
800M	410	490	515	32	29	49	55	58	60	33	57	64	67	69
900M	435	520	555	28	26	46	52	55	57	30	53	60	64	66
1000M	455	545	585	26	23	43	50	54	56	27	50	58	62	64

387

Table A-4 Aluminum Conductors

Wire Size	Ampere Rating			Direct Current	Volt Loss (See Explanation Above)									
					Three-Phase—60 Hz, Lagging Power Factor					Single-Phase—60 Hz, Lagging Power Factor				
	Type T, TW (60°C Wire)	Type RH, THWN, RHW, THW (75°C Wire)	Type RHH, THHN, XHHW (90°C Wire)	(90°C Wire)	100%	90%	80%	70%	60%	100%	90%	80%	70%	60%
Aluminum Conductors in Iron Conduit														
12	15	15	15	6040	5230	4760	4260	3740	3243	6040	5500	4920	4320	3745
10	25	25	25	3800	3291	3005	2690	2380	2080	3800	3470	3110	2750	2395
8	30	40	40	2390	2070	1905	1725	1525	1330	2390	2200	1990	1760	1540
6	40	50	55	1530	1325	1238	1126	1005	890	1530	1430	1300	1160	1030
4	55	65	70	966	837	795	726	647	585	966	918	838	747	675
2	75	90	95	606	526	511	473	434	397	606	590	546	498	456
1	85	100	110	480	415	414	386	355	330	480	478	446	410	380
0	100	120	125	382	331	336	317	294	277	382	388	366	340	320
00	120	135	145	302	262	272	260	244	232	302	314	300	282	268
000	135	155	165	240	210	225	217	206	199	242	260	250	238	230
0000	155	180	185	192	168	185	182	175	173	194	214	210	202	200
250M	180	205	215	161	142	163	163	157	153	164	188	188	181	177
300M	205	230	240	134	119	141	142	140	141	137	163	164	162	163
350M	230	250	260	115	102	126	128	127	125	118	146	148	147	148
400M	250	270	290	101	91	115	120	119	122	105	133	138	137	141
500M	270	310	330	80	74	100	104	106	107	85	115	120	122	124
600M	285	340	370	67	62	88	95	98	101	72	102	110	113	117

700M	310	375	395	58	55	82	88	92	97	64	95	102	106	112
750M	320	385	405	54	52	79	85	89	94	60	91	98	103	108
800M	330	395	415	50	49	76	83	87	93	57	88	96	101	107
900M	355	425	455	45	45	72	80	83	88	52	83	92	96	102
1000M	375	445	480	40	42	68	76	81	85	48	79	88	93	98

Aluminum Conductors in Nonmagnetic Conduit

12	15	15	15	6040	5230	4750	4250	3720	3217	6040	5490	4900	4300	3715
10	25	25	25	3800	3290	3000	2680	2360	2040	3800	3460	3100	2730	2360
8	30	40	40	2390	2070	1900	1701	1501	1304	2390	2190	1970	1740	1510
6	40	50	55	1530	1325	1230	1110	990	866	1530	1420	1280	1140	1000
4	55	65	70	966	837	787	715	641	570	966	908	826	740	656
2	75	90	95	606	525	504	462	419	378	606	580	534	484	435
1	85	100	110	480	416	405	376	343	312	480	468	434	396	360
0	100	120	125	382	331	328	307	282	258	382	378	354	326	299
00	115	135	145	302	262	265	251	232	216	302	306	290	268	249
000	130	155	165	240	208	217	206	175	164	240	250	238	202	189
0000	155	180	185	192	166	177	171	161	154	192	204	197	186	176
250M	170	205	215	161	139	153	151	144	138	161	177	174	166	159
300M	190	230	240	134	116	133	132	127	125	134	153	152	147	144
350M	210	250	260	115	100	117	117	114	114	115	135	135	132	131
400M	225	270	290	101	87	106	107	106	105	101	122	124	122	121
500M	260	310	330	80	70	89	92	92	91	81	103	106	106	105
600M	285	340	370	67	59	79	83	83	83	68	91	96	96	96
700M	310	375	395	58	50	71	76	78	82	58	82	88	90	94
750M	320	385	405	54	48	68	73	75	76	55	79	84	87	88
800M	330	395	415	50	44	66	71	73	74	51	76	82	84	86
900M	355	425	455	45	40	61	67	69	71	46	70	77	80	82
1000M	375	445	480	40	36	57	63	66	67	41	66	73	76	78

Open Wiring

The volt loss for open-wiring installations depends on the separation between conductors. The volt loss is approximately equal to that for conductors in nonmagnetic conduit. Three-phase figures are average for the three phases.

Index

Index